PROFESSION OF CONSCIENCE

PROFESSION OF CONSCIENCE

THE MAKING AND MEANING OF LIFE-SCIENCES LIBERALISM

Robert Hunt Sprinkle

PRINCETON UNIVERSITY PRESS

PRINCETON, NEW JERSEY

PUBLISHED BY PRINCETON UNIVERSITY PRESS, 41 WILLIAM STREET,
PRINCETON, NEW JERSEY 08540
IN THE UNITED KINGDOM: PRINCETON UNIVERSITY PRESS,
CHICHESTER, WEST SUSSEX

LIBRARY OF CONGRESS CATALOGING-IN-PUBLICATION DATA

SPRINKLE, ROBERT HUNT.
PROFESSION OF CONSCIENCE : THE MAKING AND MEANING OF
LIFE-SCIENCES LIBERALISM / ROBERT HUNT SPRINKLE.
P. CM.
INCLUDES BIBLIOGRAPHICAL REFERENCES AND INDEX.
ISBN 0-691-03365-X
1. LIFE SCIENCES—SOCIAL ASPECTS. 2. LIFE SCIENCES—POLITICAL ASPECTS.
3. LIFE SCIENCES—MORAL AND ETHICAL ASPECTS. I. TITLE.

QH333.S68 1994 174′.957—DC20 94-9662

THIS BOOK HAS BEEN COMPOSED IN BASKERVILLE

PRINCETON UNIVERSITY PRESS BOOKS ARE PRINTED ON ACID-FREE PAPER
AND MEET THE GUIDELINES FOR PERMANENCE AND DURABILITY OF THE
COMMITTEE ON PRODUCTION GUIDELINES FOR BOOK LONGEVITY OF THE
COUNCIL ON LIBRARY RESOURCES

PRINTED IN THE UNITED STATES OF AMERICA

10 9 8 7 6 5 4 3 2 1

TO THE MEMORY OF MY FATHER,
WHO COULD NOT TELL A LIE;

FOR MY MOTHER,
WHO FIRST CAN DO NO HARM; AND

FOR MY BROTHER AND HIS WIFE,
WHO CANNOT BE DEFEATED

CONTENTS

ACKNOWLEDGMENTS

THE FOLLOWING piece of scholarship has a complex history, its spirit having first disturbed the author's adolescent mind in the fall of 1966 and having then progressively nudged his professional course away from the studies he spontaneously enjoyed and toward "learning the human species"—toward the life sciences and clinical medicine, both of which he slowly grasped and finally grew to love.

The author, in his day, was not a conspicuous coward. His myopia kept him out of Naval ROTC at Dartmouth; his draft-lottery number was 354; only the "doctor draft" could have touched him. But he was not politically active when he should have been, and he has long felt a debt to those who were.

To friends and colleagues and teachers in medicine he sends greetings and gratitude. To friends and colleagues and teachers at Princeton and elsewhere for their more immediate help he says thanks by name: Dick Ullman, Dick Falk, Don Stokes, Bob Gilpin, Frank von Hippel, Michael Doyle, John Ikenberry, Jimmy Trussell, Peter Brown, Dick Zaner, Ed Pellegrino, Lynn Nyhart, Larry Brown, and Malcolm DeBevoise, editor and champion. To the former citizens, Khmer and Vietnamese, of what was for far too long the Site 2 refugee camp on the Thai-Cambodian border and to Nakada Nanako at Tokyo University he says what is barely safe to feel and can never fully be written down.

The author's progress toward his second doctorate was supported generously by a MacArthur Foundation Fellowship in International Peace and Security, awarded by the Social Science Research Council, New York, and announced in a densely cryptic TELEX delivered to a dusty shirt pocket in Aranyaprathet, Prachinburi Province, Thailand. His efforts were also supported generously—indeed, cheerfully—by the Woodrow Wilson School of Public and International Affairs, Princeton University. Thanks be to both.

And thanks unending to his magnificent parents, who opened the world for their sons.

R. H. S.
Thanksgiving 1993

PROFESSION OF CONSCIENCE

I

A HISTORY OF CONVICTIONS

I WAS just about to lock the door of my hotel room and go to bed when there was a knock on the door and there stood a Russian officer and a young Russian civilian. I had expected something of this sort ever since the President signed the terms of unconditional surrender and the Russians landed a token occupation force in New York.

Here was an odd beginning for a paper featured in a scholarly journal. The speaker was about to describe his own arrest. Not on an unspecified charge, as in Franz Kafka's *Trial*, or on an unfounded charge, as in Joseph Stalin's Terror. The speaker knew what the accusation against him would be, and he knew he was guilty:

> The officer handed me something that looked like a warrant and said I was under arrest as a war criminal on the basis of my activities during the Second World War in connection with the atomic bomb. There was a car waiting outside and they told me that they were going to take me to the Brookhaven National Laboratory on Long Island. Apparently, they were rounding up all the scientists who had ever worked in the field of atomic energy.

There was new guilt as well as old. In the Third World War as in the Second, reasons of state had entrapped a rising class of scientists:

> Once we were in the car the young man introduced himself and told me that he was a physicist. . . . He was obviously very eager to talk. He told me that he and the other Russian scientists were all exceedingly sorry that the strain of the virus which had been used [to defeat the United States] had killed such a disproportionately large number of children. It was very fortunate, he said, that the first attack was limited to New Jersey and that the early cessation of hostilities made attacks of larger scope unnecessary. According to plan—so he said—stocks of this virus were merely held in reserve for an emergency. Another virus differing by five further mutational steps had been in the stage of pilot plant production, and it was this improved virus that was meant to be used in case of war. It would not affect children at all and would kill predominantly men between twenty and forty.

The author described his own interrogation at Brookhaven and then the proceedings of an international war-crimes tribunal begun a month later at Lake Success. His crimes he conceded; his culpability, he claimed, had been mitigated by subsequent actions. But his prosecutor did not agree. Week by week came evidence incriminating scientists and American officials, retired or deposed: Secretary of War Henry L. Stimson, President Harry S Truman, Secretary of State James F. Byrnes. Condemnations were anticipated, but commutations were thought possible for all (except, perhaps, for Byrnes, who had once written about removing the Red Army from Eastern Germany by force). Yet, before any sentence could be passed, all indictments suddenly—and for a most extraordinary reason—were dropped:

> The first Russian appeal for help reached the United States Public Health Service one week after Mr. Byrnes rested his defense.
>
> Just what happened will never be known with certainty. This much is clear, that the vast quantities of vaccine which the Russians held in readiness to safeguard their own population against the virus were absolutely without any effect. . . . Since the engineer in charge of the production plant at Omsk perished in the disorders which broke out after over half the children of the town had died, and since all records . . . were destroyed in the fire, we shall never know just what had gone wrong.

"My Trial as a War Criminal" appeared in *The University of Chicago Law Review* in the autumn of 1949.[1] The journal's editors had wanted to publish the story in June 1948, just before the start of Truman's reelection campaign. But the author had thought an early date intemperate, and publication was delayed: ". . . political tensions would have made it difficult for the reader to be receptive to the particular treatment of the subject adopted in the article."[2]

The author of this fantasy was Leo Szilard, the Hungarian physicist famous for having conceived the feasibility of a nuclear chain reaction while crossing a London street in 1933; famous for "having tried [successfully] to induce the United States government to take up the development of atomic energy [and weaponry] in a meeting held October 21, 1939,"[3] as charged, accurately enough, by his fictional Russian prosecutor; famous, along with Enrico Fermi, for having built in a Chicago squash court in 1942 the world's first nuclear reactor. Szilard, more than anyone else, had invented the nuclear arms race, yet he had come to argue desperately against the decision to use atomic weaponry in war—a decision made by his "fellow defendants," Stimson, Truman, and Byrnes, a decision applauded by many of his fellow physicists.

In the 1930s and 1940s, physics, the first "super science," had undergone conceptual and technological transformations that had so changed its immediate value to states and its foreseeable value to society and to society's corporations that it ceased to be its own political and ethical master. Professionalism had set no defense; it raised no general alarm and lit no path to safety. Physicists, along with their colleagues in chemistry and engineering, awoke to moral bewilderment. They were scholars; they were citizens; they were patriots; they were faithful employees of academic and duly constituted civil, military, and corporate authorities. True, their modal attitudes seem to have been those of similarly cosmopolitan intellectuals; notably, many had been early and aggressive enemies of the New Order in Europe. But, manifestly, they had no special obligation, no *profession-specific* obligation, to preserve human life or to advance human welfare—the welfare of individual human beings, however aggregated. Ahead lay a future spent refining the first-ever practical means of same-day multinational murder-suicide; a future spent imagining all manner of munitions for land, sea, air, and space; a future spent promoting a short-term energy source proverbial for long-term toxicity.

Personally, physicists often came to regret their own conduct as morally incoherent. Collectively, they found that "Physics," their profession, had nothing particular to say either to them or to their societies on matters of politics and ethics. "Physics" had no built-in preventive against misuse; it had no political tradition, no ethical center.

After the atomic bombing of Japan, Szilard, of all people, became a biologist—and a distinguished one. In 1946, he was made Professor of Biophysics at the University of Chicago's Institute of Radiobiology and Biophysics. Szilard may have been attracted by the comparative political innocence and ethical clarity of the life sciences. Like physics *before* Szilard—and before Teller and Fermi and Heisenberg and Lawrence and Oppenheimer and Sakharov—biology explained creation without endangering it. Biology served the human family without dividing it and without arming its divisions one against the other, except rarely in fact and occasionally in fiction. And, even in fiction, biology came to the aid of any and all, including the miscreant biowarriors of Omsk; it was directly to the United States Public Health Service, not to the State Department, that the Russians appealed *in extremis.*

Yet, as forewarned in Szilard's projection of battles fought by virologists, biology's political innocence and ethical simplicity, long underexamined, would also soon enough be outgrown. In the 1970s and 1980s, the life sciences underwent transformations resembling in quality and potentially in consequence those that had earlier revolutionized and distorted physics.

These transformations followed two specific events, the first ancient, random, and uncelebrated, the second Promethean.

Long ago in bacteria, or in their ancestors, there arose by chance certain transcribable nucleic acid sequences—genes—whose long protein end products folded into novel electrochemically active shapes. These proteins shared an odd property. Each "recognized" and stuck to one of many nucleic acid sequences foreign to bacteria; each then disrupted its "target" sequence, always "cutting" in the same way and at the same place. These newly evolved bacterial proteins were "restriction enzymes," and the foreign sequences whose expression they "restricted" were the genes of viruses called bacteriophages. Bacteria able to restrict the genetic expression of the viruses invading them had a selective advantage, and their progeny naturally came to fill a huge evolutionary space.

Later—say, hundreds of millions of years later—life scientists isolated from common colonic bacteria the first of many restriction enzymes to be characterized. When used in concert with other techniques and with other enzymes of different classes, these newly available molecular tools imparted abilities far beyond those of the bacteria in which they had evolved, far beyond those of any living thing in nature. Life scientists could now open and close genetic sequences and alter, delete, reproduce, rearrange, or recombine genes *intentionally*. They had stolen fire.

Their success did not go unappreciated. Or unpunished. Life scientists could now change in wonderful or grotesque ways the manner in which individual members of *Homo sapiens* propagated their own or other species, produced their food, cured their diseases, organized their industries, and waged their wars.

Admiration was spontaneous and sincere but was mixed with other sentiments: apprehension, resentment, even contempt, and, from some quarters, envy. Life scientists habituated to the high-heat environment of research ethics and clinical ethics were relatively well prepared for apprehension, resentment, and contempt; others, unacclimatized, adjusted less comfortably. But few were well prepared for the last hostile sentiment: envy. Or for the grand entrepreneurial temptations that the envy of others inspired quickly in their own community of researchers and clinicians.

There is now reason to predict that life scientists will continually (and, perhaps, increasingly) find themselves unnerved by their own technology *and* pressed by others to use it—or to withhold it—in ways that may be unwise or unfair or unethical.

Some will not notice this new pressure. Others will not mind it. Still others will be intrigued by it and will learn skillfully to apply it to their own, or to their former, colleagues. But many will want fervently to resist it, and these will surely seek early tactical moral refuge in their professional identity. In other words, in their profession-specific political-ethical tradition.

What will they find there?

Not much at the moment. Not "on the shelf." Not nearly as much as leaders of (and apologists for) states and corporations and "society" have long found in the political-ethical traditions best suited to advance and excuse their interests and actions.

All of which leaves life scientists in a position not unlike that held by physicists half a century ago.

It is the aim of this book to provide for life scientists of the future what was and remains unavailable to physicists of the past and present: a profession-specific political ethics critically compared in theory and events with its inevitable philosophical antagonists.

Why has the political ethics of the life sciences not previously and variously been described?

Self-evidently, profession-specific ethical traditions adhere first to distinctive professional practices: attorneys protect interests, confessors keep confidences, physicians relieve suffering. But the breadth of these traditions varies.

Attorneys are officers of courts; they are often authors and interpreters of legislation; they are frequently civil servants and, in republics, candidates for office. Though their proper conduct in these extended roles may not uniquely be defined by codes of political ethics, attorneys have offered copiously their own thoughts on the propriety of certain political acts and, within different polities, on the moral wisdom of political philosophies and programs. Within the legal profession, there are *many* political-ethical traditions.

Clergy and theologians, likewise, have often been political actors and political-ethical commentators. For centuries in the Old and New Worlds and in many modern states—Algeria, Egypt, El Salvador, Germany, India, Iran, Israel, Peru, the Philippines, Poland, Russia, South Africa, Turkey, the United States, Vietnam—the political-ethical views of clergy and theologians have been prominent and problematical. Some of these views have achieved a degree of regularity: liberation theology, the Islamic-law movement, Christian-fundamentalist activism. Even more so than with the legal profession, clerics and theologians have produced and still follow not one but *many* political-ethical traditions.

Yet the life-sciences ethical literature, as currently defined and interpreted, has spoken only infrequently, and almost always reactively, to political matters. Political ethics has been an unusual, and rarely more than an *implicit*, topic. True, *clinical* ethics has often usefully been set, and has even been set systematically,[4] in political-secular and political-philosophical contexts: say, access-to-health-care as a peculiarly American problem or, generally, as a distributive-justice problem. But *system-*

atic political ethics in and of and for the life sciences has been a non-topic—this in contrast to the blizzards of literature in research ethics and clinical ethics that have come predictably to follow the breaking of investigational rules and technological barriers.

Why?

Are we trying to comprehend a group too disparate to share a body of experiences and opinions, too disparate to have a communal sense of right and wrong?

Possibly. Even in this book, we must use a set of designations too broad for common conversation. The term "life sciences" has to encompass biology's myriad branches, disciplines, and professional applications. "Life scientist" has to accommodate uncounted varieties, and phrases such as "life scientists and physicians" or "life scientists, physicians, and nurses" or "clinicians and researchers" have to suggest commonalities while acknowledging distinctions. The word "profession" itself, in its several forms, has to take a meaning that is exceedingly, if not excessively, general.

Alternatively, many might assume that political self-absorption has been fated by the profession's individual-centered ethical attention—classically, by its concentration on the physician-patient relationship.

But might not commitment to individual welfare just as readily be expected to have had an opposite, broadening effect?

In fact, it has.

Life scientists have demonstrably addressed political-ethical controversies, and they have done so in ways strongly influenced by the intellectual habits and ethical tendencies of their profession. Since the Nuremberg Trials following World War II, ethically principled political activism among life scientists, physicians, midwives, and nurses has been widely and, for the most part, favorably noted. Various humanitarian and internationalist agendas have been advanced by individuals and by coalitions exhibiting a range of specific commitments and preferred tactics. Poverty alleviation, civil-liberties advocacy, torture-victim rehabilitation, war prevention, war resistance (conscientious and political), war-effects and war-crimes surveillance, and refugee relief have been most celebrated. Of different character has been a self-regulatory movement, its most studied event the Asilomar Conference of molecular biologists, quietly convened in 1975, early in the restriction-enzyme era.[5]

The avoidance of nuclear war particularly has attracted life-sciences activists, "West," "East," and "South." In 1961 in the United States, Physicians for Social Responsibility (PSR) was formed,[6] and, in 1962, its early members published in the *New England Journal of Medicine* consequence projections for a multimegaton nuclear attack on Boston.[7] Bernard Lown, M.D., PSR founder, went on in 1981 to found International Physicians for the Prevention of Nuclear War (IPPNW), to which group was awarded the 1985 Nobel Prize for Peace. Lown, a cardiac electrophysiologist, had taken a broader view of his chief research interest, sudden death:

> The baton in the relay race against a nuclear Armageddon is being passed from physicist to physician. . . .
>
> Physicians do not possess special knowledge of [*sic*] direct experience in atomic matters. They do, however, have unique expertise in [related] areas. . . . The physician's movement is compelled by a growing conviction that nuclear war is the number one health threat and perhaps constitutes the final epidemic for which the only remedy is prevention. . . .
>
> A world movement of physicians is now emerging. . . . The guarded optimism pervading the physician's movement derives from an abiding faith in the concept that what humanity creates, humanity can control.[8]

The prevention of a nuclear war (or, from the Japanese perspective, a *second* nuclear war) had automatic public-health appeal worldwide throughout the era of "superpower" hostility. Witness the *Lancet*'s 1988 series of twelve papers by IPPNW supporters.[9] The last of these twelve was a paper entitled "Medical Education and Nuclear War." It included the following section:

> **THE PHYSICIAN'S OATH**
>
> Medical oaths and codes are intended to state the fundamental values of the profession. It seems reasonable then to include in these oaths a clear commitment to the medical responsibility to educate about the consequences of nuclear war. In the United States it has been proposed that medical students and physicians recognise this responsibility by adding to the oath taken on graduation a statement such as, "Recognizing the danger that nuclear weaponry represents for mankind I promise to work for peace and the prevention of nuclear war." The authors of this proposal noted that in 1983 such a sentence was added to the oath taken by graduating physicians in the Soviet Union. At its annual convention in 1984 the American Medical Student Association passed a resolution urging its members to include in their graduation oath a statement about the physician's responsibility to work for the prevention of nuclear war. IPPNW through its affiliated national organisations has pressed for such changes to physicians' oaths worldwide, but we do not know how widespread this practice has become.[10]

An interesting and revealing suggestion. At once, bold and tentative; at once, worldly and unaware. Just nuclear war? Not conventional, chemical, or biological war? Not the use of hunger as a weapon? Not the interdiction of medical trade as a military strategy? Not repression, racism, or torture? Not economic or educational or intergenerational injustice?

Not health-care insufficiency for the poor, the shunned, the foreign? Not health-insurance unavailability for those most likely to become unhealthy? Not pharmaceutical profiteering? Not the patenting of naturally occurring proteins or "invented" animals? Not the copyrighting of nucleic acid sequences?

Unfair questions? The cited paper treated nuclear war, a prospect deserving continual and concentrated forethought, even after the "Soviet disunion." The paper's authors surely did not intend to exclude from professional consideration any other problem.

But why amend an ancient oath, modern issue by modern issue, when the larger political-ethical tradition to which that oath belongs—but for which it is only the starting point—speaks to "issues-of-the-moment" categorically and timelessly?

Perhaps because these authors, like the many activists (and the many more nonactivists) they may represent, have only implicitly been aware that they are, in fact, heirs to that tradition. And because being only *implicitly* aware of a tradition approximates being without it.

Luckily, the researchers and clinicians composing the next "super science" need not proceed dimly in this way, feeling along in the half-light of moral intuition. Their forebears served them—served us all—better than that. A profession-specific political-ethical standard does not prescribe opinion; it does not predetermine conduct; and it does not guarantee discipline. But it can be instructive for initiates and practitioners alike, and it should be encouraging to all those who understand its development, respect its motives, and share its goals.

"Life-sciences liberalism" is such a standard. We will follow it from conception in Hellenic and Roman imperial times, through birth and baptism in the Scientific Revolution, then through a naïvely optimistic adolescence in the nineteenth and early twentieth centuries, and finally into a self-conscious maturity, solemnized at the Nuremberg Trials and tested ever more subtly since. Throughout, the protagonist will be a set of ideas.

In this book, with apology though not without reason, Western and Anglo-American biases will be unmistakable. True enough, the modern life sciences stand upon a Western epistemological foundation; the life-sciences literature is disproportionately composed of the output of Anglo-American universities; the language of modern science is English. But our biases are not explained, nor could they be defended, by citation of these facts; unarguably, contributions to the life sciences have been and continue to be worldwide, and full induction of non-Western societies into the "science order" promises, hopefully, a final erosion of parochialism. The explanation is more particular, for it was in the Western and Anglo-American experiences, and there only, that modernization of method in the life sciences and liberalization of ethics in politics reached their most critical points at the same time. In the same community. Even, to exaggerate a bit, in the same career.

We will find the major themes of this tradition, ancient and modern, best set forth in the epistemological and political-ethical writings of a leading seventeenth-century scientific revolutionary and much-consulted physician, the Englishman John Locke, whose political philosophy, written secretly and published anonymously, became the most widely admired since antiquity. We will find this philosophy more brightly reflective of its author's medical heritage, scientific thought, and clinical experience than is commonly recognized. We will also find that, in the Western and Anglo-American communities more evidently than elsewhere, the "Lockean Person," young or old, male or female, was to become in the mid-twentieth century the idealized subject of the life sciences, the idealized beneficiary of life-sciences ethics—investigational, clinical, political.

For reasons of peculiarity and continuity, then, Western and Anglo-American biases will prevail, and the centerpiece of our exposition will be Locke's work, significantly reassessed. This centrality may seem a contrivance, since Locke's work is crucial to so much else as well. But it is not a contrivance, as will be demonstrated in a six-part "proof."

Contributions to life-sciences liberalism have been many, if not always manifest. None has passed exclusively through the prism of Lockean political ethics, though most have gone closely by it. In the flow of exposition, we will consider the independent yet parallel work of diverse thinkers and activists: Hippocrates, Celsus, Scribonius Largus, Francis Bacon, Thomas Browne, Locke's chief mentor Robert Boyle, Thomas Percival, Claude Bernard, the "medical democrats" and the "sanitarians," Jean-Henri Dunant, Clara Barton, Hans Driesch, and, especially, William Osler, preeminent clinical scientist of the modern era and latter-day champion of Browne and Locke. Entirely outside this stream in gentle defiance stood a last great contributor: Albert Schweitzer, a non-Lockean life-sciences liberal whose "reverence for life" turned the key of professional moral purpose.

Life-sciences liberalism may be the most presentable exemplar of its philosophical class, but it has not been the only well-respected member. From 1858 until 1947, or between Osler and Schweitzer in our essay's elastic chronology, three "nonliberal alternatives" came to prominence and then to grief. *Zellenstaat* or "cell-state" theory, social Darwinism, and the eugenics movement—each was a catastrophic innovation in the political ethics of the life sciences; each was a product of science misread, each a creation of the "best" people with the "best" intentions; each proved synergistic with the other two, and all were effectively delegitimized by the Nuremberg Tribunal in a first-and-definitive fusion of research ethics, clinical ethics, and political ethics.

This exposition will not be a history of behavior. The life-sciences professions have attracted or produced many men and women of high purpose and great skill but also countless scoundrels and incompetents. This will not be their story, though it will be, in a sense, written for them, or for their successors, good and bad.

Nor will this exposition contribute to the growing literature of sociobiology, an intellectually formidable discipline whose epiphanal lessons—among the most important in modern science—have sometimes been overextrapolated to the scale of a demi-philosophy, drawing political-ethical inspiration[11] from the behavior of eusocial animals, such as ants, from the outcomes of complex game-theoretic population-genetics models, and from the remote promise of "[a] genetically accurate and hence completely fair code of ethics."[12]

How human we are. And how *un*surprising to find ourselves looking to a new source of history—molecular genetics—for a modern moral authority: evolutionary sociobiology. We will find there much that will explain why we act as we do, but will we find there *anything* that explains how we *should* act? Will we find moral authority in our chromosomes? Does the phrase "genetic accuracy" have a meaning in or for ethics, and, if achieved, could "genetic accuracy" make a code of ethics "completely fair"? The extinction of noble human traits may have been as frequent as their conservation, yet the urge to judge nobility itself evolved in *Homo sapiens*, if not in other thinking species, in perfect ignorance of classical and molecular genetics. Through sociobiology we get to know ourselves better, but it is through moral philosophy that we get to decide if we like who we are. Molecular genetics is historiography, and the human genome shows the fortune, the chance, the scars of our evolution; it is the grave goods of our species, the best and the worst our ancestors could manage. The human genome is ancient starlight; we cannot read our future there, only our past.

Admittedly, life-sciences liberalism owes its greatest single debt to the natural-law tradition; Locke learned from nature and he argued "sociobiologically" from nature—but, typically, from *human* nature itself, other-animal natures being cited analogously but not being probed for covert moral revelations. Since the seventeenth century, the soul of life-sciences liberalism has been tolerance of individual and collective variation, tolerance of eccentricity, not behavioral convergence and not ethical determinism.

If its reduction to a single aphorism were required, life-sciences liberalism might be this: "The life-scientist serves all best by serving each alone." There is safety in this rule, as will be argued, but there is also frustration. If he must serve all by serving each, how can the life-sciences liberal ever work for a "greater good," except incrementally? How can he serve—how *should* he serve—*politically*?

We will examine this problem by considering practical and potential effects of two modern "alternatives," only one of them "nonliberal" but neither of them derived from readings of the life sciences, neither of them refutable by biological argumentation, both of them vigorous though inveterate, both of them persuasive and provocative, both of them challenging philosophically and tempting professionally. First, the ethics of the state, the corporation, the institution, consolidated as the ethics of political realism. And, second, the ethics of public health, discussed as "domestic" utilitarianism.

It is service to the individual, seen or unseen, present or future, that makes moral sense of research and practice. It is service to aggregates of individuals—state, corporation, institution, society—which has led again and again to trouble. Throughout five chapters, we will see this service to the "greater good" for what it has sometimes become.

But *not* for what it must always be. There *is* for the life scientist a "greater" good—made up exclusively of "lesser" ones.

II

FROM FIRST PROBLEMS TO THE EDGE OF MODERNITY

LACKING a widely accepted orderly method, save rhetoric, for the settling of its arguments, premodern Western science was ineluctably controversial and epistemologically divergent. At its best, it was ambitiously descriptive, erratically empirical, and cautiously interpretive. At its second best, it was speculative, tendentious, teleological, and declaratory—even when experimental.[1] At its third best, it was consciously dogmatic and elaborately scholastic.

Within the "life sciences" (natural philosophy, botany, zoology, medicine, surgery), the most instructive chronic controversy involved two inconstant "groups" of physicians: those hoping to understand the living world chiefly through the fashioning of postulates, and those willing to accept whatever understanding accompanied the successful management of clinical problems. Surely, neither group was homogeneous, and, in its written record, neither adhered consistently to the opinions through whose abstraction it would come to be remembered; but no surprise in that. Leaders among the latter group, the "clinical-experience group," were the Hippocratics, famous for observational sophistication, for medicinal restraint and dietetic enthusiasm, for wound management, for closed orthopedic manipulations, and for "first, doing no harm." These Hippocratics were practical and, we might guess, unstylish men. One of them, writing *On Ancient Medicine*, declared, "[M]y view is, first, that all that philosophers and physicians have said or written on natural science no more pertains to medicine than to painting. I also hold that clear knowledge about natural science can be acquired from medicine and from no other source."[2]

Practice, not theory; observation, not postulation; proof, not pedantry. Here and elsewhere in the Hippocratic corpus was a rambunctious attitude with great long-term promise but low initial salability. The Hippocratics had little to offer but accurate prognostication, and they knew it. "Philosophers and physicians," by comparison, had everything—and an explanation for everything—to offer. It would not be till the Scientific Revolution in the seventeenth century that the Hippocratics would finally begin to prevail, the importance of clinical objectivity and the testability of hypotheses informing a newly empirical philosophy of science and, with it, as we shall see, a new political ethics.

Ancient Greek philosophers, most impressively Aristotle, treated biology, health, and medicine as scientific subjects, but they also regularly treated these topics in other ways. In several of his later works, Plato encouraged statesmen to lead their peoples to safety and to civic virtue just as free-born physicians led their patients-and-peers to bodily health: by teaching, persuading, encouraging, coaxing.[3] In a less flattering simile, Aristotle recommended that statesmen adapt ideal constitutions to nonideal peoples, just as "the trainer and the gymnastic master" adapted exercises to different body types, athletic skills, and competitive ambitions; "we notice this happening similarly in regard to medicine, and ship-building, and the making of clothes, and every other craft."[4]

Medical practitioners were, no doubt, craftsmen, as Aristotle the physician's son acknowledged, and with rare exceptions they held places low in the intellectual and social taxonomies. Image and self-image and imagination aside, of course, ancient physicians still had to tend the sick and wounded, comfort the dying, counsel the surviving, extract fees, engage consultants, and regulate their own behavior and that of their fellows. So, they did have a need for one special type of philosophy: professional ethics. And that they generated themselves, but in trace amounts.

The Hippocratic Oath, so called, is the best remembered artifact of ancient medical ethics in the West. Like much once attributed to Hippocrates of Cos (born *circa* 460, B.C.), the Oath is an item of uncertain and presumably multiple authorship. It may have been more Pythagorean than Hippocratic[5] and was largely a pledge of professional solidarity. Yet, for all its oddities, the Oath did require its adherents to treat their patients with respect and with some measure of beneficence.

The Oath spoke as much to the physician's private temptations as it did to the patient's individual needs, and it said nothing about the needs of society. Service to a good greater than the individual good was not addressed one way or another, and abuse of individual welfare for collective benefit was not clearly proscribed. Oath or no Oath, the ethical value of even atrociously cruel acts remained open to discussion.[6]

The Roman medical encyclopedist, Celsus (born *circa* 25 B.C.), when he commented on experimental vivisections performed by Herophilus and Erasistratus on conscious condemned prisoners in Alexandria, noted the justifications offered by the "dogmatic" school: "They hold that Herophilus and Erasistratus did this in the best way by far, when they laid open men whilst alive—criminals received out of prison from the kings—and whilst these were still breathing, observed parts which beforehand nature had concealed, their position, colour, shape, size, arrangement, hardness, softness, smoothness, relation, processes and depressions of each, and whether any part is inserted into or is received into another."[7]

Celsus went on to cite the several heuristic advantages of vivisection over postmortem study. Then, still representing the dogmatists' position, he wrote, "Nor is it, as most people say, cruel that in the execution of criminals, and but a few of them, we should seek remedies for innocent people of all future ages."[8]

Did Celsus share this proto-utilitarian view? It was not shared by the "empirical" school, whose members condemned such investigations—but for reasons as much epistemological as ethical. There was no certainty that the organs of a dying man were equivalent to those of a living man and, anyway, the physician could easily avoid becoming a "medical murderer" and a "cut-throat" simply by observing the anatomy and physiology exposed fortuitously in the wars, crimes, accidents, and sporting events of the day. Thus, as related by Celsus in *De Medicina,* the empiricists argued that "an observant practitioner learns . . . not when slaughtering, but whilst striving for health; and he learns in the course of a work of mercy, what others would come to know by means of dire cruelty."[9]

Did Celsus, the encyclopedist, have his own position? He did:

> Therefore, to return to what I myself propound, I am of [the] opinion that the Art of Medicine ought to be rational, but [ought also] to draw instruction from evident causes, all obscure ones being rejected from the practice of the Art, although not from the practitioner's study. But to lay open the bodies of men whilst still alive is as cruel as it is needless; [to lay open those] of the dead is a necessity for learners, who should know positions and relations, which the dead body exhibits better than does a living and wounded man. As for the remainder, which can only be learnt from the living, actual practice will demonstrate it in the course of treating the wounded in a somewhat slower yet much milder way.[10]

Celsus found convincing the dogmatists' claim that experimentation—postmortem dissection if not vivisection—was necessary for the student. But on the question of research ethics he endorsed the empiricists' argument and, likewise, their Hippocratic claim that experience, not untested theory, was the proper guide to practice. The argument against cruelty in science may at first seem the only one here touching our subject, but the other two arguments—*against* untested theory and *for* learning from experience—were also to prove important, as we shall see in the works of Sir Francis Bacon (1561–1626), Robert Boyle (1627–1691), Thomas Sydenham (1624–1689), and, ultimately, John Locke (1632–1704).

Less famous than the Hippocratic Oath and *De Medicina* was the clinical and ethical manifesto of a Roman physician of the first century A.D., Scribonius Largus, perhaps the son of a slave and perhaps personal physician to Claudius Caesar during the reconquest of Britain. Scribonius saw medicine as a fundamentally ethical enterprise whose true objective was patient welfare, and he saw patient welfare transcending loyalty even to the state.[11]

Scribonius may have been the first to present a political-ethical conclusion based, however loosely, on a medical teaching. Drugs of putative value were regularly shunned by therapeutic nihilists of Scribonius's era, to the frustration of physicians willing to use whatever seemed efficacious. In his book, *On Remedies*, Scribonius advocated the employment of certain drug-based therapies in whose efficacy he believed, and he criticized physicians who chose to forgo "the utility of drugs" through "ignorance of their craft" or, worse, "because they are subject to bias, an evil that should be despised in every living creature, especially physicians."

Scribonius went on directly from this condemnation of *bias* to a declaration of *beneficence*, generalizing his argument from practice to politics: "All gods and men should hate the doctor whose heart lacks compassion and the spirit of human kindness. These very qualities, after all, preclude the physician, bound by the sacred oath of medicine, from giving a harmful drug even to an enemy—yet the physician will attack the same enemy, when occasion demands, in his role as a soldier and good citizen. Medicine, however, does not measure a man's worth according to his wealth or character, but freely offers its help to all who seek it, and never threatens to harm anyone."[12]

Physicians *as physicians* could not kill for the state; even the emperor's own physician, as we may figure him to have been, could not kill for the empire.

Scribonius realized that a view of medicine as a profession, as a calling, as a fundamentally ethical enterprise, was not a Hippocratic view. But, he suggested, the Oath did have its humanitarian parts and a liberal reading of them could accommodate a broader ethical ambition:[13] "In truth, [Hippocrates] believed it to be of the utmost importance that each and every physician preserve the name and honor of medicine by working conscientiously, even reverently, in accordance with the maxim he himself set down: 'Medicine is the science of healing, not of harming.' If, while aiding the suffering, the doctor does not concentrate his whole being on following this ideal in every way, then he does not truly practice the compassion he promises."[14]

Scribonius was not a scientist, not a philosopher, just a physician. He may have been influenced by the stoic philosophy Cicero had presented in his book, *On Duties*,[15] but, even if he was, he went beyond it. By contrast, antiquity's most important commentator on the Hippocratic corpus, Galen (A.D. 129–199), the physician who dominated Roman medicine a century later and then dazzled Islamic and European medicine—save Paracelsus (1493–1541), the "medical Luther"—for fifteen centuries and more after that, failed even to mention the Hippocratic Oath in his works now surviving, and he noted Scribonius only on technical matters; he had no use for his ethics.[16] The humanitarianism and universalism suggested by the Oath and expanded by Scribonius must have rested within a range of acceptable medical-ethical views but may still have been exceptional.

More significant than these few treatments of medical ethics were other classical works by physicians and natural philosophers and technologists and even poets, works that exuded the liberal self-confidence of people anxious to base their views not on revelation or authority but on their own observations or on the observations of others they judged reliable.

Classed with such as these would have been the Hippocratic authors of *On Ancient Medicine* and *Epidemics I* and *III*—among whom may have been the real Hippocrates of Cos, student of Democritus—as well as the "methodic" physician and skeptical commentator, Sextus Empiricus. So, too, the atomists Leucippus, Democritus, and Epicurus, as revealed by Diogenes Laertius and as championed by the poet Lucretius; the indispensable Euclid and Archimedes; the heliocentric astronomer Aristarchus of Samos; Vitruvius of *De Architectura*; and Ovid, agnostic mythologist of the *Metamorphoses*. Many would not have been included, though, despite monumental achievements. Pythagoras, the mystical mathematician, would not have qualified, nor Empedocles, the charismatic pathophysiologist, nor, perhaps, even Galen, premier investigator of his day, a genius strangely—tragically—restrained by intellectual authority, even by authority of his own invention or adaptation, a genius easily tempted to set discovery aside for rhetorical advantage.[17]

And Aristotle? If there was ever to be a political-ethical tradition in the life sciences, should it not have begun with Aristotle, "father of biology" and author of the *Politics* and the *Nicomachean Ethics*?

Yes, the tradition should have begun with Aristotle, and in obvious ways it did. It was Aristotle who "[i]n biology . . . [spoke] for the first time the language of modern science," and who "seems to have been first and foremost a biologist," and whose "natural history studies influenced profoundly his sociology, his psychology, and his philosophy," and "whose language is our language, whose method and problems are our own"—as recalled by an ever-collegial William Osler, speaking in 1919.[18]

But Aristotle's contribution to the tradition would be an ambiguous one.

At the start of the *Politics*, Aristotle declared his intention to study the state by dissection, and he spoke more than metaphorically of civic ontogenesis. In the work as a whole, he examined, ordered, and related all known species of constitutions, producing, almost overtly, a comparative anatomy and physiology of politics. Here, surely, was Aristotle as founder.

But here also was Aristotle as *con*founder, for in the same opening arguments was an error in reasoning—teleology—that distorted Aristotelian science and Aristotelian political ethics alike, an error not clearly identified till the Scientific Revolution, an error not widely understood till the latter half of the nineteenth century and not sufficiently feared even then (or, say most scientists and many philosophers, even now): "nature makes nothing [for more than one purpose], but one thing for one purpose."[19] To know purpose by use—this was to know, wholly without warrant, that wings existed to allow flight, rather than to know, empirically and reliably, that many winged creatures could fly. This was also to know that king ruled commoner, master ruled slave, man ruled woman, and Greek ruled barbarian *naturally*. Men, to say nothing of men and women, were not by nature interchangeable, and they were not *by rights* equivalent. Life-sciences liberalism still lay in the far distance, but its first nonliberal alternative already lay in wait.

Centuries hence, as the West rediscovered its future during a long convalescence from the Black Death, ancient authority finally began yielding to ancient skepticism, and the Renaissance and Reformation thinkers willing to trade authority for empiricism, to trade dogma for observation, theory for experience, deduction for induction, were the thinkers noticeably ahead of their times. It would be toward the Hippocratics, the skeptics, and the atomists and away from the paralytically enthralled Aristotelians and Galenics that the Scientific Revolution would propel life scientists and, inseparably, their political ethics. Yet, as late as 1550, it was with the still-potent authority of Aristotle's "scientific" political ethics, specifically with his doctrine of natural slavery,[20] that Juan Ginés de Sepúlveda (1490–1573), translator of the *Politics*,[21] could counter Bartolomé de Las Casas (1474–1566), Dominican ex-bishop of Chiapa[22] and campaigner against the sins of the Spanish Conquest,[23] in their epic imperial debate on the rights and right uses of New World aborigines.[24] They argued to a draw, exhausting their jury with a catalogue of philosophical incompatibilities. When controversy cooled, the emperor kept his subjects, the pope kept his souls, and the conquistadors made of the aboriginal remnant an eventually thriving serfdom—while buying or stealing *indisputedly* "natural" slaves on the saddest coast of the other and older world.

III

FROM THE SCIENTIFIC ATTITUDE TO THE UNIVERSALIST SENTIMENT

SIR FRANCIS BACON, Baron of Verulam, Viscount St. Alban, Lord High Chancellor of England, progenitor of the Royal Society, reputed back-stabber, convicted bribe-taker, was no ethicist. Neither was he a life scientist. And he was certainly not a physician. He failed even to note William Harvey's revolutionary work on the circulation of blood, though Harvey attended him personally.

That said, there was no scientific revolutionary after Bacon, not in England anyway, who did not owe to this insistent yet elusive man the honor of antecedence. The debt was not technological; Bacon imagined and suggested and proposed and tinkered but did not invent. The debt, rather, was epistemological. Bacon showed how to know.

Did he also show *why* to know? Did he say to what ends his scientific method could morally be applied? And, if he commented on political ethics at all or on the political ethics of science especially, did he say the same to every audience, or was his advice ambiguous—or even disingenuous?

On the one hand, Bacon was a political realist. Most Lords High Chancellor are political realists. He often quoted milder parts of Machiavelli with approval. He thought the balance-of-power concept the only "generall Rule" of international relations, and he approved the preemptive strike as a lawful response to "a just Feare, of an Imminent danger."[1] He advised "above all, for *Empire* and *Greatnesse*, it importeth most; That a Nation doe professe Armes, as their principall Honour, Study, and Occupation."[2] And, more ominously, he advised "Nations, that pretend to *Greatnesse*" to be sensitive to minor provocations, so as not to overlook "just Occasions (as may be pretended) of Warre." Like a human being, a state needed exercise to maintain its health; Bacon prescribed foreign war as the ideal outdoor activity for the body politic.[3]

On the other hand, Bacon was a liberal. His views on religious toleration were distinctly liberal for his day, as Locke's would be also. Though, like Thomas Hobbes (1588–1679), he listed "*Innovation* in *Religion*" first among the causes of sedition,[4] Bacon did not favor forced religious conformity: "Concerning the *Meanes of procuring Unity*; Men must beware, that in the Procuring, or Muniting, of *Religious Unity*, they doe not Dissolve and Deface the Lawes of Charity, and of humane Society."[5]

In his extensive writings on the theory of just war, Bacon deemphasized, though he did not eliminate, the role of religious difference in excusing military action. There were uncivilized societies, piratical nations, and cruel empires against which Christian war could be waged, but their listing should be well justified, and the list would not be a long one. In 1622, a year after he fell from power, in the summarizing speech of a dialogue on the ethics of "Holy War,"[6] Bacon expressed an opinion not always shared by former heads of government: "It is a great error, and a narrowness or straitness of mind, if any man think that nations have nothing to do one with another, except there be either an union in sovereignty or a conjunction in pacts or leagues. There are other bands of society, and implicit confederations. . . . But above all . . . , there is the supreme and indissoluble consanguinity and society between men in general: of which the heathen poet (whom the apostle calls to witness) saith, *We are all his generation.*"[7]

Of course, it is not as a politician that Bacon is chiefly remembered, but as a philosopher of science, as a campaigner against "narrowness or straitness of mind." He ridiculed the technological and industrial sterility of authoritarian natural philosophy and urged the abandonment of the overly deductive methods of "scientific" logicians. Shunning received wisdom, Bacon recommended his replacement method as foolproof: induction from observation followed by experimentation designed to "falsify" incorrect tentative conclusions. Bacon's scientific method was a sensational advance for his day. He did not get things exactly right, however, giving too small a role to theoretical thinking. Claude Bernard (1813–1878), French anatomist and experimental physiologist, thought Bacon "a great genius" but "not a man of science" and doubted that he understood "the mechanism of the experimental method."[8] Mathematical philosophers such as Bertrand Russell (1872–1970) and philosophers of social science such as Karl Popper (born 1902) have taken him unkindly to task for his errors. But many of Bacon's perceived errors[9] have in fact been misperceived[10] by his critics, and Bacon's place in the philosophical pantheon now seems secure.

For originality, Bacon deserved high marks. But, to an extent inadequately acknowledged by his more admiring biographers, Bacon's vision was a reflected one, and the changes he advocated were already under way. For instance, William Gilbert, an English physician, published one of the first results of the scientific method, *De Magnete*, in 1600, well before the scientific method had been "invented" by Bacon. Gilbert was criticized by Bacon, but not for doing bad science and not for doing science before being told how to do it by Bacon, as is often charged by Bacon's detractors. Gilbert was criticized for drawing unwarranted conclusions from his observations, for lessening the value of his own work.[11] We look back on Gilbert's book generously, excusing its inferential hyperextension as merely the mark of its time. Bacon read Gilbert's work as a contemporary, praised it for what it had added to a still-narrow base of reliable data, and condemned it for the illegitimate speculation by which it would surely mislead the next investigator interested in the same topics. This was Bacon as peer reviewer, equal parts advocate and adversary. And it was Bacon in the context of his age.

Why was the Lord High Chancellor so interested in science and technology? Intellectual ambition may be an adequate answer, but it has not been the only answer suggested. Puritan millenarianism, centering on the fourth verse of the twelfth chapter of the Book of Daniel, predicted that "knowledge shall be increased" as humanity regained the dominion over nature that had been compromised at the Fall. The start of the Millenium was widely predicted for sometime in the seventeenth century, and efforts to advance learning seemed prudential to believers wanting to act "elected," as they hoped the Final Judgment would prove them to have been. Was Bacon one of these millenarians? He did quote the Daniel verse in advancing and defending his recommendations,[12] and there is good reason to think he was a man of sincere religious beliefs. But the "great instauration" he foresaw was to be man-made—the beginning of progress, not its culmination.

Was he trying to expand his personal authority? Much hostility to Bacon has come from the notion that he advocated the use of science to gain and wield political power. This idea has been made more plausible by his own personal conduct and, as we have seen, by several unattractive passages preserved in his essays. Similarly to his later discredit, when recommending to King James his book, *The Advancement of Learning*, Bacon took pains to reassure his sovereign that learning would not make his kingdom unwarlike. Furthermore, both openly and in the utopian parable, *New Atlantis*,[13] Bacon proposed the creation of research institutions whose common task it would be to collect data for unspecified purposes. Such a proposal from an admirer of Machiavelli has been looked on disapprovingly by some, and in the twentieth century—the century of totalitarianism—it has been called "chilling."[14]

But Bacon was not a convincing authoritarian, and he was not a scientific nationalist. True, he urged his king to found and to fund institutions and investigators whose output would advance society and state. But he simultaneously urged the internationalization of that output. He wanted "more intelligence mutual between the universities of Europe than now there is"; he wanted to nurture "a fraternity in learning and illumination."[15]

Bacon is remembered ambivalently for the aphorism, "Knowledge is power." Sometimes[16] a bit unfairly,[17] but not always: "Human knowledge and human power meet in one; for where the cause is not known the effect cannot be produced. Nature to be commanded must be obeyed."[18] Such an equation of knowledge and power-over-nature was innocent enough in a preindustrial age, and its ethical intention must have been slight. The scientist as malicious manipulator is not found here or elsewhere in Bacon's works.

Indeed, Bacon's overt treatment of the knowledge-and-power issue set for the scientist a high standard of faithful and self-denying stewardship:

> But yet evermore it must be remembered that the least part of knowledge passed to man by this so large a charter from God must be subject to that use for which God hath granted it; which is the benefit and relief of the state and society of man; for otherwise all manner of knowledge becometh malign and serpentine. . . .
>
> And therefore it is not the pleasure of curiosity, nor the quiet of resolution, nor the raising of the spirit, nor victory of wit, nor faculty of speech, nor lucre of profession, nor ambition of honor or fame, nor inablement for business, that are the true ends of knowledge; some of these being more worthy than other, though all inferior and degenerate: but it is a restitution and reinvesting (in great part) of man to the sovereignty and power . . . which he had in his first state of creation. . . . And therefore knowledge that tendeth but to satisfaction is but as a courtesan, which is for pleasure and not for fruit or generation.[19]

Though not published until 1734, this passage had become available to scholars and disciples immediately following Bacon's death.[20] Bacon's assessment "[o]f the limits and end of knowledge"[21] could not easily have been confused: knowledge was to be used to benefit humankind; its use for more particular goals was immoral.

But was humanity morally competent? Bacon himself had been tested repeatedly in public life and, by accounts, had failed. In high places, Bacon reflected, "There is License to doe Good, and Evill; whereof the latter is a Curse; For in Evill, the best condition is, not to will; The Second, not to Can. But Power to doe Good, is the true and lawfull End of Aspiring."[22]

Admittedly, much of what Bacon wrote on knowledge, power, and moral choice could be described as aspirational. Yet, Bacon was a historical optimist,[23] and, unlike some of the political realists with whom he has been classed, he believed that humanity did have an inborn moral competence: "The Inclination to *Goodnesse*, is imprinted deepely in the Nature of Man: In so much, that if it issue not towards Men, it will take unto Other Living Creatures: As it is seen in the Turks, a Cruell People, who neverthelesse, are kinde to Beasts, and give Almes to Dogs, and Birds."[24]

Bacon did not share the political realists' cynical opinion of human nature, nor did he share their pessimism about the human future. He saw both good and evil in humanity but he thought the "inclination to goodness" could be fostered—fostered specifically by the material products of Bacon's own scientific method. And, on this topic, where he may have felt less open to a charge of hypocrisy, he spoke idealistically, as in the Preface to *The Great Instauration*: "Lastly, I would address one general admonition to all; that they consider what are the true ends of knowledge, and that they seek it not either for pleasure of the mind, or for contention, or for superiority to others, or for profit, or fame, or power, or any of these inferior things; but for the benefit and use of life; and that they perfect and govern it in charity. For it was from lust of power that the angels fell, from lust of knowledge that man fell; but of charity there can be no excess, neither did angel or man ever come in danger by it."[25]

By the judgment of eighteenth-century commentators, Bacon's influence was torrential; by the judgment of nineteenth-century commentators, it was narrow and quick to recede.[26] Today, there is little doubt about Bacon's importance, but there is debate about his culpability. Some, regretting and expecting misuses of modern science, have imagined Baconian philosophy directing the human species toward self-enslavement[27] or, through "the new biology," toward "voluntary self-degradation and dehumanization."[28] Such criticisms may seem overwrought, or, when directed individually against a long-dead man, unfair. But they have been offered conscientiously. Indeed, they manifest in extremity the less fevered concern motivating the present book.

But our immediate interests are elsewhere directed. Our objects now are the careers of two Baconian ideas: first, that plain, unprejudiced observation is the key to understanding; and, second, that science must be used only for the good of humankind.

The first of these Baconian ideas was not new. It had had its ancient advocates, particularly among Hippocratic and methodist physicians. But much different, much less irreverent principles had animated medieval scholastics; change, progress, correction, improvement—though evident in Aristotle[29]—had not been emphasized in scholastic interpretations of the classical legacy. Bacon chose not only to transmit the more pliable, more liberal, more liberating parts of the past; he also chose to transmute them, to advance their evolution more quickly and less randomly than it would otherwise have been advanced. Bacon urged the application of his method to problems in science and technology, but there was no reason it could not be adapted to problems in ethics and politics. And it was so adapted, though in its nonexperimental guise, Bacon's scientific method becoming the Baconian scientific attitude. Avoidance of premature decision became avoidance of prejudice; willingness to follow wherever observation and inference might lead became tolerance; readiness to acknowledge the value of ideas and accomplishments on merit rather than on source became universalism and internationalism. Certainly, this adaptation of the scientific attitude did not fully permeate Western society; and it incompletely penetrated even the small community most likely to express it: scientists themselves. Still, this attitude did affect the lives and thoughts of many members of that community and, to the extent that it reentered broader society through their works and writings, it affected much more besides.

The second of these Baconian ideas—that science must be used only for the good of humankind—also had had its antecedents, as we have seen. But never before had this idea been tied to a philosophical system so well developed and so indisputably "scientific." This was philosophy speaking ethically *because* it was speaking comprehensively; this was science speaking ethically *because* it was speaking from itself and of itself. The Lord High Chancellor could have portrayed science speaking differently: science for military engineering, science for economic hegemony, science for state strength. Now and then in Bacon's *Essayes* and in *The Advancement of Learning* and in his history of Henry VII's reign,[30] the reader almost suspects the man's intentions. But, judging his whole output, his intentions are hard to fault. Bacon's sanctified science would have had difficulty serving nonhumanitarian, nonuniversalistic ends. Had Bacon wished really to serve such ends, he could not have sanctified science in the first place; he would have had to sanctify something else, such as the state, as Hobbes would sanctify the state a generation later. Bacon served both the state and science, and he took pains to make science compatible with the state. But, like Scribonius, he did not place science in service to the state. And, if he had tried to do so, if he had tried to become a Machiavelli of science, it is not at all sure that the members of the scientific community to which he and his published works spoke would have paid much attention.

Consider Thomas Browne (1605–1682).

Here was a famous physician of sound education, a knight by a stroke more of luck than of sword,[31] a man whose interest in a call to scientific arms would likely have been nil. Most of our exposition will parallel the growth of skepticism, experimentalism, and experientialism, but the pleasantly unavoidable Doctor Browne exhibited an epistemology as Platonic as it was Baconian.

Around 1634, just having taken his medical doctorate, Browne wrote some warm-hearted, half-mystical thoughts on the reconciliation of science and religion and on the coexistence of one religion with another. Shown to a friend who in turn showed them to a circle of friends, these writings were by 1642 being sold in two unauthorized editions under the title *Religio Medici*, "The Religion of a Physician." This was the year the English Civil War began, and Browne, a well-established practitioner and a Royalist, thought it wise to "correct" his work for an authorized version, which appeared in 1643 under the same title. *Religio Medici* was remarkably widely read in England and in continental Europe; by the time the Stuarts were restored, ten authorized editions had been consumed; many more would follow.

In an era of religious warfare, Browne declared himself zealously adherent to the established faith of his own country but simultaneously tolerant of those who, by misfortune, as he saw it, had been taught differently: Protestant sectarians, Catholics, Jews, Mohammedans. The key to peace was reservation of moral judgment, and reservation of moral judgment was a simple, logical function of relative circumstance: "No man can justly censure or condemne another, because indeed no man truely knowes another. . . . Further, no man can judge another, because no man knowes himself."[32] Charity was a duty, persecution evil and foolish, national hatred and ethnic prejudice pointless.

Browne, he said, could hate none of his fellow human beings: "[N]ationall repugnances doe not touch me, nor doe I behold with prejudice the *French, Italian, Spaniard,* or *Dutch.*"[33] But he could hate the political phenomenon—collective action—through which his fellows committed their greatest sins: "If there be any among those common objects of hatred I doe contemne and laugh at, it is that great enemy of reason, vertue and religion, the multitude, that numerous piece of monstrosity, which taken asunder seeme men, and the reasonable creatures of God; but confused together, make but one great beast, & a monstrosity more prodigious than Hydra; it is no breach of Charity to call these fooles."[34]

Charity itself was not to be defined narrowly, not to be seen only as the giving of material goods. And, for a physician, charity had a contractarian dimension: "To this (as calling my selfe a Scholler) I am obliged by the duty of my condition, I make not therefore my head a grave, but a treasure of knowledge; I intend no Monopoly, but a Community in learning; I study not for my owne sake onely, but for theirs that study not for themselves."[35]

As a further aid to those who studied not for themselves, Browne went on to publish, in 1646, *Pseudodoxia Epidemica*, a book of amazing length by its fifth edition.[36] In *The Advancement of Learning*, Bacon had urged the compilation of "a calendar of popular errors," a catalogue of the "dross and vanity" that "weakened" and "imbased" human knowledge.[37] *Pseudodoxia Epidemica* was Browne's response: an epidemiology of false doctrine, an assault on the barriers to scientific revolution, a revolution in which Browne, an irrepressible provincial medical experimenter, saw himself a participant. Ancient authority, unskeptical reasoning, rumor, and gullibility were the targets.

How revealing of the man and the times, then, to note that, in 1664, the year he was elected a Fellow of the Royal College of Physicians,[38] Browne testified as an expert witness for the prosecution in the trial of two "witches" at Bury St. Edmunds.[39] Universalism was not yet for women.

What should we make of Sir Thomas Browne? Was he trying to do for Protestantism and the science of the future what Aquinas had tried to do for Catholicism and the science of the past? Was Browne trying to reconcile inspiration and investigation? Did he make an original contribution to the political-ethical development of the life sciences, or did he simply reflect that development, without having the intellectual scope to understand it? And, particularly, was he tolerant by temperament or tolerant by training? And, even if the latter, how tolerant was he, really,[40] and how tolerant was his profession?

Our certainties are few and indirect. *Religio Medici* came to be widely read and highly regarded by centuries of physicians, its twin messages of professional benevolence and religious, national, and ethnic tolerance recognized unambiguously as aspects of the life-sciences attitude. Boyle read and often quoted the book.[41] Locke owned the fourth edition.[42] Sir William Petty (1623–1687)—doctor of medicine and professor of anatomy at Oxford, physician to Cromwell's army of Irish occupation, founding fellow of the Royal Society, father of demography, inventor of "political arithmetick," "political anatomy," and empirical economics[43]—listed *Religio Medici* first among the three books "most esteemed and generally cried up for wit in the world," as recorded in coffeehouse conversation by Samuel Pepys (1633–1703).[44] Sir William Osler (1849–1919) counted *Religio Medici* his favorite book[45] and spent twenty years searching out its surviving editions.[46] He lectured on and wrote about Sir Thomas and included *Religio Medici* with the Bible and Shakespeare among the ten titles in his "Bed-side Library for Medical Students."[47] His much-traveled companion copy, placed on his coffin in the Chapel of Christ Church, Oxford, New Year's Day, 1920,[48] is now displayed at the Osler Library, McGill University.[49]

If our exposition were to end here, after Hippocrates, Celsus, Scribonius, Bacon, and Browne, we would have unearthed little more than a substratum of convictions. True enough, ethics, including political ethics, must be founded on emotional choice: *for* compassion and *against* pain, *for* equality and *against* privilege. But ethics left at this level is easily overshadowed by higher structures built with rational arguments: compassion may be admirable, but it is also inefficient and often dangerous; equality may seem fair, but it unjustly frustrates the strong and the talented; the protection of individual human rights is important, but it is less important than the maximization of social welfare.

The political ethics of the life sciences, as late as the 1660s, lay almost entirely unassembled. It may even have been unenvisioned. If our exposition were to end here, we would have nothing useful to say.

IV

FROM THE SCIENTIFIC REVOLUTION TO THE LIBERAL EXPECTATION

EASILY A MATCH for Bacon and altogether more formidable than Browne was John Locke, a "man of parts" so excellent and various that even his finest biographers have failed to describe him whole. Locke the political philosopher, Locke the epistemologist, Locke the scientist, Locke the physician—all these might as well have been different men, judging from the scholarly record. But they were not different men.

John Locke was the great synthesizing figure of the Enlightenment, recapitulating and then stimulating the intellectual development of his age. He cannot properly be understood as anything else. Yet his surviving twin reputations as epistemologist and as political philosopher have so dominated his twin reputations as scientist and as physician that these less remembered twins are usually assumed to have died in infancy. But, in fact, they were born first, they prospered, and they taught their younger siblings much of what later made them famous.

That said, if John Locke, no less, is to be identified as principal architect of the political ethics of the life sciences, then deep within Locke's political ethics we will have to find weight-bearing members indisputably of life-sciences origin. And, correspondingly, when examining the political ethics of the life sciences, we will have to find Locke's mark displayed in latter-day additions to the structure *and* at sites of integral repair. Failing these tests, Locke-in-the-life-sciences will have to be dismissed as a case of casual visitation.

Our exposition of the political ethics of the life sciences now requires a Lockean excursion. We shall organize it into six investigations:

1. Was Locke just an observer of the Scientific Revolution, or was he a co-conspirator? Did he abandon unverifiable authority independently or, at least, enthusiastically? Was he significantly influenced by pioneering experimentalists and empiricists, and did he influence any of them in return? Were there political-ethical standards being proposed or displayed by scientists other than Locke, and, if so, was Locke either naturally in agreement with them or ultimately persuaded by them?

2. Was Locke legitimately a physician by the standards of his day? If a physician, was he a rationalist, an experimentalist, or an empiricist? And did he in any way affect ambient practice standards or, failing that, the habits of individual clinicians?

3. Did Locke's scientific orientation and clinical experience affect the empirical epistemology of *An Essay concerning Human Understanding*? Did the empiricism of the *Essay* appeal to later life scientists and physicians?

4. Did Locke's empiricism affect the "liberal" political ethics of the *Letter on Toleration* and the *Two Treatises of Government*? If so, was Locke's liberalism the inevitable product of his physicianhood?

5. Did "liberalism" enter the political-ethical tradition of the life sciences after Locke, and, if so, did that liberalism persist into the "modern" era? If liberalism did persist, did life scientists and physicians ever associate it with Locke?

A sixth investigation, one examining "additions" and "repairs," will be implicit in later chapters.

Locke and the Scientific Revolution

In 1647, his sixteenth year, John Locke, son of a Puritan lawyer of Somerset, entered the Westminster School. It may have been, as Maurice Cranston has suggested, that Locke's first liberalizing influence was the right-wing Royalism of Westminster's vigorous headmaster, Richard Busby.[1] Puritan thesis, Royalist antithesis, liberal synthesis would have been the dialectic of this transition. We may assume, and, indeed, we know in some detail, that men and women of particular political and religious beliefs did affect Locke's philosophy. Busby may have been the first of them.

But our attention in this study is demanded by a different category of liberalizing influence, a fully interactive influence, dynamic and collegial. It, too, can be traced from the Westminster years, but it was not personified by a headmaster or recorded in an ancient book. At Westminster, Locke met among his schoolmates John Mapletoft,[2] Walter Needham,[3] and Richard Lower,[4] each to become a distinguished physician, each to become one of Locke's many scientific friends and collaborators.

Richard Lower went up to Christ Church, Oxford, in 1649[5] and was among the first of its students to study chemistry.[6] Locke took up a scholarship at Christ Church in 1652.[7] Though his curriculum was initially classical and was expected eventually to be clerical, Locke began immediately to show an interest in science. The earliest of Locke's medical notebooks dates from the first months of his Oxford years, and it was to Lower that many of its entries were attributed. Through Lower, Locke became a member of the "Oxford Experimental Philosophical Clubbe" and eventually assisted the experiments of Thomas Willis (1621–1675), the neuroanatomist and neurophysiologist; Robert Boyle, the pneumatic physicist, "father" of modern chemistry,[8] pioneering physiologist, and moralist;[9] and Lower himself, later famous as a cardiopulmonary anatomist and physiologist, as a pioneer in the transfusion of blood, and as "vindicator"[10] of the experimental methods of Harvey, Willis, and others. Among the many experiments Locke recorded in his first medical notebook were even some of his own, including a confirmation of Harvey's demonstration of the circulation of the blood and simple investigations into reflex neurological activity in a prepared frog.[11]

Locke was elected to a senior studentship in 1658 and served as a classics don.[12] At the end of 1660, he was elected Lecturer in Greek and two years later Lecturer in Rhetoric. By 1660, the Stuarts restored, Locke had become a functioning Monarchist, as had many disenchanted Puritans. He was also a Hobbesian. Even his religious views were intolerant,[13] despite attending "Lockean" sermons given by the soon-to-be-replaced dean of Christ Church Cathedral, John Owen.[14]

Christ Church accommodated fifty-five studentships for churchmen or aspirants. There were only five nonclerical studentships, and only two of the five were in medicine.[15] By 1663, Locke was expected to take holy orders.[16] He avoided the church temporarily by accepting the Censorship in Moral Philosophy and by taking a minor diplomatic post in Brandenburg. He avoided the church permanently by deciding to become a physician.[17]

Clerical life and livelihood were insecure in England in the 1660s, especially for Nonconformists, and the passive, "nonmedical" interpretation of Locke's professional distraction credibly cites "discouragement"[18] among aspiring ministers. But it less convincingly includes Locke among them. Actually, judging from his notebooks, Locke's decision for medicine was long overdue by 1666. Natural science had been his greatest interest for fourteen years, and, since 1660, he had been working closely with Boyle.[19] Beginning in that same year, feeling his ignorance of medicinal plants and knowing the gross inadequacies of available authorities, Locke began compiling his own botanical textbook; by 1664, he had classified about 1,600 specimens.[20] In 1663, he studied chemistry under Boyle's visiting German protégé, Peter Stahl.[21] In 1663 and 1664, he attended Willis's Sedleian Lectures in Natural Philosophy and studied Willis's earlier lectures from Lower's notes. Locke's Latin transcriptions of these thirty-four lectures on thirty-eight topics have given historians by far their best record of seventeenth-century medical teaching.[22]

Locke's relationship with Robert Boyle was complex and formative. It was Boyle who co-founded the "Invisible College," a loose association of Baconian enthusiasts "invisible" for geographical reasons, not for concealment.[23] Boyle's list of guiding principles for the Invisible College recalled Bacon closely: "charity," "universal good-will," intolerance of "narrow-mindedness," insistence that members take "the whole body of mankind for their care," and insistence that they enter economic enterprise only for "the good they may do with it." "Universal good-will" could be overapplied, however; one social action to which the Invisible College had been pledged during the English Civil War was the provision of saltpeter to the New Model Army.[24]

The Invisible College seems to have been centered in Oxford during Boyle's residence there; its Oxford identity was probably the "Experimental Philosophical Clubbe" of which Boyle, Robert Hooke, Lower, Stahl, Willis, Christopher Wren, and Locke were members.[25] This "College" and "clubbe" evolved into the Royal Society of London for Improving of Natural Knowledge,[26] to which Locke was elected in 1668, aged thirty-six years.[27]

Now best remembered for his pressure-volume "Law" of 1661, Boyle in his day was known not only as a chemist but also as a theoretical and experimental physiologist. His opinions were sought and respected in medical cases, though he was a gentleman-scientist, seventh son of the First Earl of Cork, and not by training a physician. He wrote extensively on medical topics, and, partly to increase the availability of affordable agents among the laboring classes, he collected, evaluated, and disseminated medicinal "recipes." Recent archival scholarship shows that he worked on a book attacking Galenism and then stopped short of publication, apparently to avoid affronting practitioners he knew to be in error.[28] For his myriad acknowledged contributions to medicine, Oxford made him Doctor of Physic, *honoris causa*, in 1665.[29] It was to Boyle that Lower dedicated his 1665 *Vindicatio* justifying Baconian experimentation in anatomy and physiology and condemning unverified assumption.[30] It was Boyle who persuaded Thomas Sydenham to begin his studies of London epidemics in 1661, and it was to Boyle that Sydenham's *Methodus Curandi Febres* was dedicated five years later. The clinical method displayed in that little book on fevers so impressed Locke that, when he went down to London in 1667, he arranged an introduction to Sydenham, probably through his friend Mapletoft, who had gone on from Westminster to read medicine at Cambridge;[31] Locke soon displaced Mapletoft as Sydenham's chief collaborator.

Locke often assisted Boyle at Oxford, learning experimental methods and learning much "iatrochemistry," or "physician's chemistry," in the process. Boyle's neo-atomistic corpuscular theory of matter was avidly adopted by Locke, whose philosophical reasoning it permeated; *An Essay concerning Human Understanding* was heavily indebted to *The Sceptical Chymist* of 1661 and to *The Origin of Forms and Qualities* of 1666. Locke and Boyle corresponded for decades and collaborated repeatedly in experiments. It was at Locke's request[32] that Boyle undertook the most important of his physiological studies, *Memoirs for the Natural History of Humane Blood,* published in 1683 with a dedication "to the very Ingenious and Learned Doctor J. L."[33] It was to Locke and Isaac Newton that Boyle bequeathed his last best guess for the transmutation of mercury to gold. It was Locke who co-executed Boyle's will in 1691.[34] And it was Locke who edited and largely rewrote Boyle's *General History of the Air,* published posthumously in 1692;[35] many of the meteorological data contained therein were from "A Register kept by Mr Locke" over a period of seventeen years.[36]

A parting question on Locke's relationship with Robert Boyle: was Boyle responsible for Locke's move away from Hobbes? We have noted that Locke was a Hobbesian of sorts in 1660 and that Locke began working with Boyle in that same year. We have also begun to see that Boyle's influence on Locke was profound *and* multidisciplinary. But these observations alone place us far short of an answer. And also, perhaps, far off the mark; it is to other candidates that Locke's liberal metamorphosis, which cannot be dated before 1666, has traditionally been credited, yet never with good evidence and never with much confidence.[37]

Let us consider the following assertion: the suppression of his more Hobbesian habits and opinions and the overlaying or substitution of Boylean skepticism, empiricism, and liberality marked *the* strategic intellectual transition in Locke's life. A fair statement? Certainly not if another mentor had in fact been more influential. Or if Locke's future variation on natural law were seen as a refinement of Hobbes's work, rather than as an alternative to it; some scholars have favored "refinement"—and a few might substitute "vandalism"—but Locke's choice of method, a choice he displayed elaborately in the *Essay,* argues compellingly for "alternative." That said, our trial assertion would again be false if Hobbes and Boyle—or "Hobbesianism" and "Boyleanism"—had not been intellectually antagonistic. If Hobbes and Boyle had somehow been compatible, then Locke's transition from the one to the other could have been a progression, not a reversal of field.

Judging from Hobbes's *Leviathan* and from Locke's references to Boyle in the *Essay,* Hobbes and Boyle *were* philosophically, politically, and ethically incompatible. But this is a judgment requiring critical analysis of Locke's mature work. It does not tell us whether Locke, in effect, had to "choose" between Hobbes and Boyle in the early 1660s.

Hobbes saw himself as a natural philosopher. He had met Bacon and

Galileo Galilei (1564–1642) and, while in Paris sitting out the Long Parliament and the Civil War, René Descartes (1596–1650) and Pierre Gassendi (1592–1655); with Petty, his much younger fellow student, Hobbes had studied optics and read Vesalius, and he had tutored the exiled Stuart heir in mathematics, especially, we may guess, in Euclidean geometry. But Hobbes was not a scientist. In *Leviathan*, published in 1651, he had drawn deductions *from* physical postulates—incorrect ones, unfortunately—*through* human nature *to* politics. *De Corpore* of 1655 covered some of the same ground. Hobbes's worldview was inconsistent with Boyle's empiricism, and Hobbes said so. Hobbes and Boyle, it is usually forgotten, were lively disputants.[38]

In 1661, Hobbes published *Dialogus Physicus de Natura Aeris*, a work unavailable in English until 1985, when Simon Schaffer presented his rendering: *A Physical Dialogue, or a Conjecture about the Nature of the Air taken up from Experiments recently made in London at Gresham College.*[39] In this tract, Hobbes challenged, both technically and epistemologically, Boyle's air-pump experiments, described just the year before, 1660, but already famous. Hobbes said Boyle's work was not philosophy; it was inadequately concerned with cause, too much concerned with observation. Boyle responded quickly and at length with *An Examen of the greatest part of Mr. Hobbes's Dialogus Physicus de Natura Aeris*; Hobbes was "a writer of politics" whose "name may with some readers give his arguments an efficacy, which their own nature could not confer on them."[40]

Locke could hardly have missed either the stridency or the significance of the Hobbes-Boyle debate. Long before Locke had started to write philosophy—to *revolutionize* philosophy by making it less concerned with cause, more concerned with observation—he had good reason, at a personal level maybe pressing reason, to turn away from Hobbes, to turn away from what would become political realism (and, more immediately, political anathema)[41] and to turn toward Boyle and true science—*and*, arguably, political liberalism. Some of Boyle's strongest views concerned the detrimental effects of private interest on the distribution and *re*distribution of the benefits of science; he complained that medicinals were often flagrantly and greedily overpriced, and he hoped the "recipes" he recommended would prove not just safe and efficacious but also affordable by working folk and the poor, for whom medical expense meant ruin as surely as did untreated illness.[42] At any rate, clearly, Locke picked Boyle. And, almost as clearly, he never reconsidered his choice.

Locke, then, was fully involved in the Scientific Revolution. He did abandon unverifiable authority both independently and collegially. He was significantly influenced by pioneering experimentalists, such as Willis, Lower, and Boyle, and by empiricists, such as Boyle again and particularly, as we shall see next, Sydenham. Did he influence any of these men in return? He probably did influence Boyle to some extent; he affected his choice of investigations, and he certainly supplied him data. Locke's influence on Sydenham we shall assess later.

Locke as a Physician

Many healers prospered in Locke's day with the thinnest of scientific and technical abilities; a man of Locke's general brilliance and charm might have grown rich as a complete charlatan had he so wished. But several facts suggest that, by 1666, the year he formally switched careers, Locke knew more contemporary medical science than his past identity as a classicist surely implied to those who did not know him well.

First, Locke was quickly able to enter into partnership with an established Oxford physician, David Thomas.[43] At Boyle's suggestion,[44] Locke established with Thomas and another physician a small iatrochemical laboratory. We now know that in this laboratory in 1666 Locke concocted or compounded nearly every medicinal substance in the seventeenth-century pharmacopoeia.[45]

Second, when the chronically ailing Anthony Ashley Cooper, an amateur experimenter in his own right, came to Oxford to drink therapeutic waters supplied by Doctor Thomas, Locke so impressed him that Cooper soon urged Locke to establish a medical practice in and for the Cooper household in London. Cooper, who would soon say, with good reason, that he owed his life to Locke's clinical judgment, was not a man easily fooled. When, as Lord Ashley, Baron of Wimbourne St. Giles, this same Cooper became Chancellor of the Exchequer, and when, as First Earl of Shaftesbury, he became Lord High Chancellor of England, Locke was still his personal physician and, more famously, his political and economic counselor.

Third, by 1667, only a year after his formal shift to medicine, Locke began working with Thomas Sydenham,[46] a physician now legendary as "the English Hippocrates," by most modern estimations the greatest clinician of his age, and, by all contemporary accounts, a man thoroughly intolerant of incompetence. It was with Locke's help that Sydenham expanded his study of febrile diseases. And it was with Sydenham's help that Locke completed his escape from intellectual prejudice.

Thomas Sydenham was a clinical revolutionary, an apostate preaching the Hippocratic method of observation, description, and limited intervention. He trusted little beyond his own experience and was correctly disdainful of most contemporary medical doctrine. The good physician helped the human body cure itself by augmenting the body's own therapeutic capacity—its *physis*, its *natura*. Iatrochemical interventions were illegitimate unless observed to be beneficial in specific syndromes. Hippocrates, Bacon, and Boyle were Sydenham's principal influences,[47] and, like them, Sydenham followed an empirical method. But, unlike Bacon and Boyle, who were experimentalists as well as empiricists, Sydenham dismissed dissection, microscopy, and physiological experimentation as wastes of time. Why speculate on ultimate cause when practical effect was still obscure? Usable knowledge of prognostics and therapeutics was the critical need and could be gained only at the bedside.

Did Sydenham find in Locke a well-prepared physician or a medical mind unspoiled because it was medically undereducated? No one trained at a university—even at Oxford, like Sydenham himself—learned much that was useful in practice. And Locke had not yet qualified even as a bachelor of medicine, let alone as a doctor of medicine. But this was not unusual for physicians of the day, and Locke, we now know, had been studying medically relevant subjects for fifteen years. It could hardly have been Locke's academic innocence that appealed to Sydenham. It was much more likely Locke's enthusiasm for Sydenham's empirical method that made this partnership an enduring success.

Until he met Sydenham, Locke had filled his notebooks not only with the faulty rationalistic speculations passed on by Willis and Lower and other Oxonians, but also with the results of their far more reliable experimental work. Willis's *Cerebri Anatome* was being prepared while Locke was at Christ Church. Lower performed the brain dissections under Willis's direction, and Christopher Wren developed original methods for preparing the tissue for study and then drafted the book's spectacular illustrations. This was work of a high order, quite different in quality from the now-humorous jumble of pseudo-physiological nonsense that filled the gaps between cogent autopsy anecdotes in the Sedleian Lectures. Locke recognized the difference between these two types of intellectual product, and, in his masterpiece on empiricism, *An Essay concerning Human Understanding*, he would make clear which he favored. Locke would retain his commitment to experimentation, despite Sydenham's misgivings.

But, still, in 1667, at Sydenham's side, Locke was ready to suppress the good with the bad, at least temporarily, and Willis and Lower disappeared from Locke's notebooks and correspondence. Sydenham may have been wrong about the value of experimentation, but Locke knew Sydenham was right about the sterility of classical medical education. It was clinical empiricism, not laboratory empiricism, that Locke wanted to learn from his new teacher. And it was clinical empiricism, even more than laboratory empiricism, that Locke would later adapt to philosophy, politics, and ethics.

Locke's work with Sydenham during the smallpox epidemics of 1667 and 1668 was demanding and, depending on the state of Locke's immunity, dangerous. But Locke's enthusiasm for the task and his admiration for Sydenham's clinical method were undiminished. A 1668 reprinting of Sydenham's book on fevers, by then known as *Observationes Medicae*, was prefaced by twenty-seven Latin couplets extolling the author's clinical method, signed "J. LOCK, A.M., Ex Aede Christi Oxon."[48] The partnership of these two men is an old and true medical legend. Collaboration was close for several years and correspondence active till Sydenham's death. Locke attended Sydenham's own son when the boy had the measles,[49] and Sydenham consulted for Locke when the chronic hydatid cyst in Lord Ashley's liver turned acutely suppurative. And so on.

We should pause briefly on the matter of Lord Ashley's liver. Though still a newcomer to clinical medicine, Locke managed Ashley's case admirably, by Osler's analysis. Locke considered the opinions of at least eight other physicians, directed a daring incision and drainage, attended a long and complicated convalescence, and finally adapted to Ashley's residual fistulous tract a silver drainage tube, "Shaftesbury's Tap" or the "Shaftesbury Spigot" of Restoration satire. His progress notes constituted the first detailed description of such a case in the history of medicine,[50] and the entire record, expert commentary folded in, would have constituted a "grand rounds" or "clinicopathological conference" if presented didactically.

Locke's attachment to Shaftesbury's household and to Shaftesbury's cyclical fortunes made maintenance of a typical practice unfeasible. By Kenneth Dewhurst's estimation, Locke had ceded most of his patients to Sydenham by 1673.[51] But this did not mean that Locke ceased being a physician intellectually or socially. In fact, much of Locke's medical career still lay before him. He "proceeded" Bachelor of Medicine on February 6, 1675, Oxford awarding him a medical license, making him a medical don, and giving him an income, but, fortuitously, not requiring his residence.[52] By the next year, Locke was in France, and Shaftesbury, for angering the Lords in Parliament, was in the Tower of London.[53]

Locke's friend Mapletoft now resumed his place as Sydenham's closest associate. When *Observationes Medicae* entered its third edition in 1676, the book's "Epistle Dedicatory" was addressed to Mapletoft, not to Locke. But it was to Locke's approbation of his controversial clinical method that Sydenham made particular reference: "You know also how thoroughly an intimate and common friend, and one who has closely and exhaustively examined the question, agrees with me as to the method that I am speaking of; a man who, in the acuteness of his intellect, in the steadiness of his judgment, in the simplicity (and by *simplicity* I mean *excellence*) of his manners, has, amongst the present generation, few equals and no superiors. This praise I may confidently attach to the name of JOHN LOCKE."[54]

In France, Locke's hosts and companions were typically physicians, and he often served as a consultant to physicians locally or by letter. The second and third "first descriptions" credited to Locke arose in this consultative capacity: *tic douloureux* or trigeminal neuralgia[55] and onychogryphosis[56] or "Locke's disease." For eighteen months of his three-and-a-half years in France, Locke resided in Montpellier, site of a great university—Rabelais had taken his medical degree there. Charles Barbeyrac, the leading physician of Montpellier, was, like Sydenham, a skeptical, Hippocratic empiricist. Barbeyrac, a Protestant, had been rejected by the university, so, despite international fame and a large practice, he was not officially a "doctor." Barbeyrac, like Hippocrates, like the "humanist" physicians of Padua,[57] like Sydenham, and like Osler, taught his students at the bedside. Locke accompanied these rounds,[58] and he and Barbeyrac became close friends.[59]

In Holland, during a later and sometimes desperate exile, Locke owed his freedom to physician friends. He also owed them no small debt of intellectual stimulation, as will be explained in our discussion of toleration. In settled times, Locke often served as an intermediary between physicians of different nationalities, helping to disseminate clinical case reports and effective treatment regimens, medical and surgical. Enlisting some of these same international correspondents, Locke, along with Doctor Charles Goodall, instigated a simultaneous survey of bills of mortality and meteorological data, the purpose being to judge the correlation of weather change with epidemic severity.[60]

By all reports, Locke's medical opinions were widely valued. When Sydenham wrote to Locke telling him he had finally decided that Peruvian bark was effective for intermittent agues, Locke, who was in France, introduced Sydenham's practice to a wide circle of physicians. Peruvian bark seems not to have been used in France prior to Locke's recommendation but soon became popular as "the English remedy" for intermittent agues.[61] "Peruvian bark" was, of course, quinine, and malaria was the most common intermittent "ague," or intermittent "acute" febrile disease. Parenthetically, it was Lower's refusal to use Peruvian bark for Charles II's ague[62] that guaranteed his subordination to Sydenham in the clinical pantheon.

For reasons academic, political, and personal, Locke was never awarded a doctorate in medicine. The professional significance of this omission has been overestimated by scholars of philosophy. Sydenham, by way of comparison, had been "created" Bachelor of Medicine by Wadham College, Oxford, in 1648 after barely a year of study[63] but did not receive his doctorate until after the 1676 edition of his book on fevers. And, even then, his degree was awarded *honoris causa* by Pembroke College, Cambridge, where he had never studied.[64]

Let us now reconsider the question of Locke's influence on the empiricist Sydenham. Osler felt the influence was clear and strong, but Osler, like most everyone before Dewhurst, assumed incorrectly that Locke had written *De Arte Medica*,[65] "an introduction to a treatise on the philosophy of medicine."[66] Dewhurst—M.D. *and* F.R.Hist.S.—has shown otherwise, as we shall see.

Patrick Romanell has argued cleverly that Locke redirected Sydenham's thinking on the pathophysiology of smallpox. The textual case rests mostly on revisions in the third edition of *Observationes Medicae* and on a scrap in the Public Record Office—Locke's "Smallpox Fragment"—in which certain comments on the primacy of observation seem to have been destined for a cooperatively planned treatise. Romanell's argument[67] is convoluted and a bit thin, but it may be right. Even if it is not right, it is still plausible. And its plausibility—a function of Locke's full inclusion in the Scientific Revolution—is the important point.

Much of the foregoing evidence was unavailable before the opening of the Lovelace Collection at the Bodleian Library in 1948[68] and before the Mellon Donations were added to the collection in 1960 and 1963.[69] Even the most optimistic Lockean medical revivalist must have been stunned at the contents: in all, about 3,500 incoming letters and 150 outgoing drafts,[70] many representing correspondence with physicians in England, Holland, France, and elsewhere; about a thousand other papers; sixteen medical notebooks; and ten journals recording many entries of medical interest, such as patient progress notes and clinical advice from colleagues. The journals dated from 1675 to 1698, long past the time many assumed Locke had ceased dabbling in medicine.[71] Included in the first Mellon Donation was Locke's herbarium, two bound volumes containing 970 of the botanical specimens collected at Oxford in the early 1660s. How revealing of the predicament of a scientist at Christ Church to note that the specimens had been mounted on the backs of student lessons turned in for grading to John Locke, classics don and Censor in Moral Philosophy.[72]

The Lovelace Collection made a complete reassessment of Locke the physician necessary; it made a rediscovery of Locke the physician-philosopher inevitable; and it made our portrayal of Locke the life-sciences liberal credible. But, really, quite a bit of evidence had been available long before 1948 and had been widely disregarded by philosophical commentators. Just why is an excellent question, still not fully answered, despite intriguing theories.[73] One reason need not be called theoretical: a lapse of careful scholarship; a famous edition of *An Essay concerning Human Understanding* will provide a sadly amusing example below.

But our present questions are more directly answerable. Was Locke legitimately a physician by the standards of his day? Yes. If a physician, was he a rationalist, an experimentalist, or an empiricist? He was both an experimentalist, a producer of data, and an empiricist, a collector of data; but he was more characteristically the latter. And did he in any way affect ambient practice standards or the habits of individual clinicians? Yes, on both counts.

Science, Medicine, and Locke's Empiricism

Locke's great *Essay* was written over a period of twenty years, but its full period of preparation might be set at thirty years, back to the start of Locke's collaboration with Boyle, whose corpuscular philosophy and observational, inductive method were among the book's fundamental themes.[74] Locke's frequent use of chemistry as a source of evidence, example, and anecdote is easily traced to Boyle and other chemists. Several of Locke's favorite, recurring metaphors, such as the lock-and-key metaphor, were taken from Boyle,[75] as was much of Locke's materialism.[76] Even Locke's nominalism, his concern for the clarity of words and ideas, had antecedents in *The Sceptical Chymist*[77] and in *The Origin of Forms and Qualities.*[78]

Also antecedent was the influence of Sydenham[79]—specifically, the empirical clinical method exhibited to generations of European physicians in Sydenham's widely read textbook, *Observationes Medicae.* Of more peculiar importance to us is *De Arte Medica,* a brief item not published anywhere till 1876[80] and then incorrectly attributed to Locke himself. Now understood to have been the work of Sydenham, it was undoubtedly known to Locke the year before he started the first draft of the *Essay.* In 1669, Locke helped Sydenham plan *De Arte Medica,* a pugnacious treatise never finished beyond its first few pages. The extant fragment is in Locke's handwriting; Sydenham, whose gouty arthritis was often debilitating, must have dictated it.

Passages in *De Arte Medica* foreshadowed major themes of the *Essay.* Ancient or modern, a "rule of practise founded upon unbiased observation" was always worthy of respect, but an unconfirmed hypothesis never, for it only "confined and narrowed mens thoughts, amused their understanding . . . , and diverted their enquirys." Deductive reasoning in medicine, as an example, had led to "endless disputes." Employing nonempirical methods led a physician to share the predicament of someone "that should walke up and downe in a thick wood overgrowne with briers and thornes with a designe to take a view and draw a map of the country." All progress had "sprung from industry and observation; true knowledg had grown first in the world by experience and rationall operations." Trying to "penetrate into the hidden causes of things" by old philosophical means had proved futile and had caused much distress; "happy discoverys" would be made only through "chance or well-designed experiments."[81]

The Locke revealed in 1690 would resemble in outline the Sydenham revealed in 1669 in *De Arte Medica.* Locke in the *Essay* was not reluctant to acknowledge his intellectual debt to "the English Hippocrates"—or to others. He modestly depicted himself in Sydenham's image of someone "in a thick wood overgrowne with briers and thornes":

> *The Commonwealth of Learning, is not at this time without Master-Builders, whose mighty Designs, in advancing the Sciences, will leave lasting Monuments to the Admiration of Posterity; But every one must not hope to be a* Boyle, *or a* Sydenham; *and in an Age that produces such Masters, as the Great* — Huygenius, *and the incomparable Mr.* Newton, *with some others of that Strain; 'tis Ambition enough to be employed as an Under-Labourer in clearing the Ground a little, and removing some of the Rubbish, that lies in the way to Knowledge.*[82]

Of course, Locke would prove a master builder as well, "British" empiricism and "Lockean" liberalism being his own "lasting monuments." It is Locke's liberal political ethics that interests us in this study, and we shall see that Locke built this edifice upon his empiricism, brick by brick. Locke was a scientist—and particularly a life scientist, a physician—writing philosophy and ethics, and his liberal arguments were ultimately based on an empirical medical model of human nature, of human emotion, of human understanding.

An Essay concerning Human Understanding is overly long and needlessly repetitious. Locke blamed the seductiveness of the subject for the book's length and the circumstances under which he wrote for its repetitions, "*[b]ut to confess the Truth,*" he explained in a letter to the reader, "*I am now too lazie, or too busie to make it shorter.*"[83] For our purposes, however, the *Essay* is fine exactly as written, since it recorded Locke's thinking over many years, and it indicated, by its redundancies, which kinds of thinking seemed to him persistently valid.

One analytical technique, "this Historical, plain Method,"[84] dominated the book, though not as exclusively as Locke said it would; Boyle's neo-atomistic corpuscular theory—a theory with a future—made a major appearance. But Locke's chief "method" was indeed "historical" and "plain": "historical" in a natural-sciences and clinical sense and "plain" in an observational and descriptive sense. Romanell, arguing from texts and fragments, has linked Locke's "historical, plain method" to Sydenham's "plain and open method."[85] But this linkage need not be discovered among archives. Familiarity with clinical thinking and clinical teaching makes it hard to overlook: when interpretation fails, describe. And, in seventeenth-century medicine, as Locke and Sydenham and their neo-Hippocratic colleagues knew so well, interpretation failed nearly all the time.

Locke's purpose in the *Essay* was from the beginning both epistemological and political. As it turned out, Locke would come to write his major works on toleration and on government only after the *Essay* was far advanced. His later arguments against certainty in religious doctrine and against governmental interference in private affairs owed much to the "scientific liberalism" first crystallized in the *Essay*:

> I suppose it may be of use, to prevail with the busy Mind of Man, to be more cautious in meddling with things exceeding its Comprehension; to stop, when it is at the utmost Extent of its Tether; and to sit down in a quiet Ignorance of those Things, which, upon Examination, are found to be beyond the reach of our Capacities. . . . If we can find out, how far the Understanding can extend its view; how far it has Faculties to attain Certainty; and in what Cases it can only judge and guess, we may learn to content ourselves with what is attainable by us in this State. . . .
>
> 'Tis of great use to the Sailor to know the length of his Line, though he cannot with it fathom all the depths of the Ocean. . . . Our Business here is not to know all things, but those which concern our Conduct. If we can find out those Measures, whereby a rational Creature put in that State, which Man is in, in this World, may, and ought to govern his Opinions, and Actions depending thereon, we need not to be troubled, that some other things escape our Knowledge.
>
> This was that which first gave the *Rise* to this Essay concerning the Understanding. For I thought that the first Step towards satisfying several Enquiries, the Mind of Man was very apt to run into, was, to take a Survey of our own Understandings, examine our own Powers, and see to what Things they were adapted. Till that was done I suspected we began at the wrong end, and in vain sought for Satisfaction in a quiet and secure Possession of Truths, that most concern'd us, whilst we let loose our Thoughts into the vast Ocean of *Being*, as if all that boundless Extent, were the natural, and undoubted Possession of our Understandings, wherein there was nothing exempt from its Decisions, or that escaped its Comprehension. Thus Men, extending their Enquiries beyond their Capacities, and letting their Thoughts wander into those depths, where they can find no sure Footing; 'tis no Wonder, that they raise Questions, and multiply Disputes, which never coming to any clear Resolution, are proper only to continue and increase their Doubts, and to confirm them at last in perfect Scepticism.[86]

Locke was to no small degree writing an essay on "Conduct," on *political* conduct, on the *irrationality* of excessive "meddling" and ill-founded "Disputes."

He set out first to explain how a particular type of idea came into the mind: that type of idea of which all were conscious and which many philosophers, notably Descartes, claimed to be "innate." Locke was considering *ideas.* He was *not* considering instincts, or what he called "natural tendencies."[87] Locke found no good argument for any idea's "innateness" or "implicitness." Rather, the seeming universality of some simple ideas was a function of inborn human capacity.[88] On this point, he was making an argument consistent not only with modern developmental neuropsychology but also with the most advanced views of his own day, as heard in Willis's Sedleian Lectures, though he did not cite Willis as a source.

Locke knew this argument against innateness and for capacity would prove politically potent:

> [I]f *these first Principles* of Knowledge and Science, *are* found *not* to be *innate, no other speculative Maxims can* (I suppose) *with better right pretend to be so.*[89] . . .
>
> Whether there be any such moral Principles, wherein all Men do agree, I appeal to any, who have been but moderately conversant in the History of Mankind, and look'd abroad beyond the Smoak of their own Chimneys. Where is that practical Truth, that is universally received without doubt or question, as it must be if innate? *Justice,* and keeping of Contracts, is that which *most Men seem to agree in.* This is a Principle, which is thought to extend it self to the Dens of Thieves, and the Confederacies of the greatest Villains; and they who have gone farthest towards the putting off of Humanity it self, keep Faith and Rules of Justice one with another. I grant that Outlaws themselves do this one amongst another: but 'tis without receiving these as the innate Laws of Nature. They practise them as Rules of convenience within their own Communities.[90] . . .
>
> I think, *there cannot any one moral Rule be propos'd, whereof a Man may not justly demand a Reason:* which would be perfectly ridiculous and absurd if they were innate, or so much as self-evident.[91]

Locke saw moral rules as the end product of what we might call "social logic," since they "plainly [depend] upon some other antecedent to them, and from which they must be deduced."[92] The ultimate antecedent might be, say, the power of God for a Christian or the punitive power of the Leviathan for "an *Hobbist*" or "the Dignity of a Man" for "one of the old *Heathen* philosophers."[93] Even the best candidate for innateness—"'*Parents preserve and cherish your Children.*'"[94]—was not always heeded. Far from being "innate," it was almost automatically stated as a command, since, "[t]o make it capable of being assented to as true, it must be reduced to some such Proposition as this: *It is the Duty of Parents to preserve their Children.* But what Duty is, cannot be understood without a Law; nor a Law be known, or supposed without a Law-maker, or without Reward and Punishment: So that it is impossible, that this, or any other practical Principle should be innate; *i.e.* be imprinted on the Mind as a Duty, without supposing the *Ideas* of God, of Law, of Obligation, of Punishment, of a Life after this, innate."[95]

We today would argue—against Locke of the *Essay* but not so much against Locke of the *Two Treatises*, as we shall see—that parental devotion is inborn and that it has co-evolved with high intelligence, prolonged offspring dependence, and low fertility in species such as *Homo sapiens.* Moreover, its inborn locus cannot be disproved, as Locke claims it can in the *Essay*,[96] by pointing out that deviant parents treat their children cruelly; variety in phenotype fuels the naturally selective evolutionary process. That said, sociobiology lay beyond Charles Darwin (1809–1882) far in the secular future—and with Lucretius even more remotely in the pagan past. Locke's understandable error here was not in ignoring "natural tendencies" entirely but in failing to place certain moral behaviors among them.

We would also have to argue, this time more fairly, that working up the causal ladder from parental motivation to duty, to law, and thence to lawmaker or to reward or punishment could not fully explain the behavior under study. A decision still had to be made to honor the lawmaker's right to make law or to forego reward or to endure punishment. If such a decision were made emotionally, then some ultimately emotional "standard" had to be postulated. If it were made rationally by a standard of self-interest, then self-interest itself had to be valued over other goods—still a problem unsolved in the *Essay*. Locke, like all philosophers, had a goal in mind, an endpoint for his argument, and he rounded some corners to get to it.

In the *Essay*, Locke's goal seems to have been this: demonstration of the role of circumstance and teaching in the growth of understanding and in the ontogeny of moral and religious views. Locke's concept of the plasticity of principles was liberating indeed in an age of religious warfare and witch hangings:

> *Doctrines*, that have been derived from no better original, than the Superstition of a Nurse, or the Authority of an old Woman; may, by length of time, and consent of Neighbours, *grow up to the dignity of Principles* in Religion or Morality. For such, who are careful (as they call it) to principle Children well, . . . instil into the unwary, and, as yet, unprejudiced Understanding, (for white Paper receives any Characters) those Doctrines they would have them retain and profess. . . .
>
> This is evidently the case of all Children and young Folk; and Custom, a

greater power than Nature, . . . it is no wonder that grown *Men* . . . should *not* seriously sit down to *examine their own Tenets*, especially when one of their Principles is, That Principles ought not to be questioned. . . . And he will be much more *afraid to question those Principles*, when he shall think them, as most Men do, the Standards set up by God in his Mind, to be the Rule and Touchstone of all other Opinions. And what can hinder him from thinking them sacred, when he finds them the earliest of all his own Thoughts, and the most reverenced by others?

It is easy to imagine, *how* by these means it comes to pass, that *Men* worship the Idols that have been set up in their Minds; grow fond of the Notions they have been long acquainted with there; and *stamp the Characters of Divinity, upon Absurdities and Errors*, become zealous Votaries to Bulls and Monkeys; and contend too, fight, and die in defence of their Opinions.[97]

Locke's "white paper" was a revolutionary refabrication of the *tabula rasa* found in Aristotle, the Stoics, the Aristotelian Peripatetics, and Thomas Aquinas (*circa* 1225–1274). Except for "natural tendencies," the mind was blank before being informed, or prejudiced, by circumstance and teaching; the mind at birth was not just unactivated, not just waiting to think innate thoughts; it was truly blank, truly uninformed and unprejudiced.

Somewhere between the old classical and scholastic views and Locke's newer perspective was the medical doctrine Locke studied at Christ Church. Willis's lecture on sensation, will, memory, judgment, and fantasy presented a transitional version of the *tabula rasa*—the "scraped tablet"—as a neurophysiological assumption consistent with facts discovered by dissection and experiment. Willis's teachings were still unacceptably rationalistic and teleological, and largely wrong, but they were opinions partially based on experience:

Imagination is caused by an impression from some external object that moves the spirits inwards and excites other spirits in the medulla oblongata into an expansive movement. These latter spirits are then variously circulated through the cerebral orbits forming different ideas. When a similar movement is repeated in these orbits memory is evoked as the cerebrum in infants is a *tabula rasa*. Phantasy arises from the convergence of various objects, as the spirits move in different ways. When imagination is evoked the expansion of spirits is repeated and the paths and orbits of the spirits are varied. . . . The substance of the cerebellum is very different from that of the cerebrum: it is firmer, and marked out with fixed orbits and paths for the spirits. There are anatomical similarities in the cerebellum of all animals, and hence, the circulation of the spirits is almost the same in all of them.[98]

Almost none of this and almost none of what followed in the balance of the lecture was correct. The comparative anatomical comment can still be appreciated, but little else. Intriguingly, the reference to the *tabula rasa* seems internally inconsistent. Locke—or Lower, from whom Locke borrowed the notes for this lecture—may have misunderstood Willis's meaning. But, more likely, Willis was reasoning just beyond the limits of his anatomically informed but physiologically biased medical system. If the infant's cerebrum was truly a *tabula rasa,* then how could "similar movement . . . repeated in these orbits" evoke memory? Perhaps the awkwardness of this part of the transcript shows a student's confusion, even disagreement. It is clear from Locke's notes that he did disagree with some of what Willis taught; to one lecture, Locke attached a comment asserting correctly that only arterial blood carried nourishment, contrary to Willis's teaching.[99] No analogous note was attached here.

In the *Essay,* Locke covered many of the topics Willis had covered in the Sedleian Lectures. Locke had far less to say in specific neurophysiological terms. However, a higher proportion of what he did say still makes sense today. By the time Locke wrote, Vesalius and then Willis, Lower, and Wren had provided a great many valid neuroanatomical data but only a few valid neurophysiological data. And, as Sydenham doubtless stressed to Locke again and again, even the neurophysiological data were pretty useless at the bedside. Almost none helped the clinician. And almost none helped the natural philosopher understand understanding.

Locke's project in the *Essay* was in large part the presentation of reliable information on human neuro*psychology*—how and why people think what they think. Neuro*psychology* could be based on observation of the whole human organism, while neuro*physiology* could be based only on experimentation in an isolated organ system. Of the two, neuro*psychology* must therefore have seemed the more retrievable to a seventeenth-century intellect:

> [H]ath not Navigation discovered, in these latter Ages, whole Nations, at the Bay of *Soldania* . . . , in *Brasil* . . . , in *Boranday* . . . , and in the *Caribee* Islands, *etc.* amongst whom there was to be found no Notion of a God, no Religion[?][100]
>
> Had you or I been born at the Bay of *Soldania,* possibly our Thoughts, and Notions, had not exceeded those brutish ones of the Hotentots that inhabit there: And had the *Virginia* King *Apochancana,* been educated in *England,* he had, perhaps, been as knowing a Divine, and as good a Mathematician, as any in it. The difference between him, and a more improved *English*-man, lying barely in this, That the exercise of his Faculties was bounded within

> the Ways, Modes, and Notions of his own Country, and never directed to any other, or farther Enquiries.[101]
>
> [O]ur minds being, at first, void of that *Idea*, which we are most concerned to have [a knowledge of God], it *is a strong presumption against all other innate Characters*. I must own, as far as I can observe, I can find none, and would be glad to be informed by any other.[102]

We should note that the manner of Locke's sequential argumentation on this point was roughly Baconian: from a tentative hypothesis, he moved to observation, then to the induction of a refined hypothesis, and then to hypothesis testing. First, he considered the innateness of some ideas. Second, he observed religious variety. Third, he *in*duced the refined hypothesis that the notion of God was not innate. Fourth, he *de*duced the tentative conclusion that there were no innate ideas of any kind and undertook to test the validity of this finding by requesting falsifying counterexamples from other observers.

The *tabula rasa* theory was a comfortable first consideration for Locke, as it had been for others before the Scientific Revolution. But Locke's investigation was unlike any of its predecessors, and its political-ethical implications were fundamental. From the epidemiology of religious variation among rational men and women, Locke was able to argue that human acts, customs, and beliefs were functions of circumstance and teaching, not functions of innate ideas, not functions of racial or national virtues.

Today, Locke's conclusion falls safely between the ethical determinism of extreme sociobiologists and the ethical *in*determinism of extreme behaviorists, both schools criticized by biologist and philosopher Ernst Mayr (born 1904), who, though not citing Locke, does seem either to have adopted Locke's view or, more likely, to have reached it independently.[103]

From circumstance and teaching was elaborated the third of Locke's three partially overlapping moral standards. The first was divine law, the second civil law, and the third was "The *Law of Opinion or Reputation*," a law universally evident to any impartial observer:

> Vertue and Vice are Names pretended, and supposed every where to stand for actions in their own nature right and wrong: And as far as they really are so applied, they so far are co-incident with the *divine Law* above mentioned. But yet, whatever is pretended, this is visible, that these Names, *Vertue* and *Vice*, in the particular instances of their application, through the several Nations and Societies of Men in the World, are constantly attributed only to such actions, as in each Country and Society are in reputation or discredit. Nor is it to be thought strange, that Men every where should give the Name of *Vertue* to those actions, which amongst them are judged praise

> worthy; and call that *Vice*, which they account blamable: Since otherwise they would condemn themselves, if they should think any thing *Right*, to which they allow'd not Commendation; any thing *Wrong*, which they let pass without Blame.[104]
>
> I think, I may say, that he, who imagines Commendation and Disgrace, not to be strong Motives on Men, to accommodate themselves to the Opinions and Rules of those, with whom they converse, seems little skill'd in the Nature, or History of Mankind: the greatest part whereof he shall find to govern themselves chiefly, if not solely, by this Law of Fashion; and so they do that, which keeps them in Reputation with their Company, little regard the Laws of God, or the Magistrate. . . . Nor is there one of ten thousand, who is stiff and insensible enough, to bear up under the constant Dislike, and Condemnation of his own Club. He must be of a strange and unusual Constitution, who can content himself, to live in constant Disgrace and Disrepute with his own particular Society. . . . This is a Burthen too heavy for humane Sufferance: And he must be made up of irreconcilable Contradictions, who can take Pleasure in Company, and yet be insensible of Contempt and Disgrace from his Companions.[105]

This was a forgiving attitude. Much or most of the world's evil was explained by societal dynamics, not by individual maliciousness.

Locke described "Unreasonableness" as a "Disease" and suggested that "Prejudice [was] a good general Name for the thing it self."[106] This particular disease was acquired by habituation, and it was therefore avoidable. To the faulty intellectual habits associated with prejudice Locke attributed much of the world's discord, discord that was tragically unnecessary. People went to great lengths to guard themselves and their children against somatic diseases, but they foolishly disregarded the threat posed by adverse psychological and psychosomatic conditioning. Locke presented several medical case histories illustrating such conditioning.[107] Ultimately, he argued, "Some such wrong and unnatural Combinations of *Ideas* will be found to establish the Irreconcilable opposition between different Sects of Philosophy and Religion. . . . [T]here must be something that blinds their Understandings, and makes them not see the falsehood of what they embrace for real Truth."[108]

"[B]y Education, Custom, and the constant din of their Party,"[109] sectarians were made politically sick. But even sectarians were rational, and they could be cured, Locke hoped, by more careful attention to the meaning of words and the use of language.

Locke's nominalism now seems a bit pedantic, but in a time of ideological violence, well before Doctor Johnson's dictionary began to stabilize the English language, an effort to clarify the link between thought and words was a serious undertaking. The substance of Locke's nominalism does not concern us, but Locke's account of an early stimulus to his thinking on the subject *does* concern us:

> I was once in a Meeting of very learned and ingenious Physicians, where by chance there arose a Question, whether any *Liquor* passed through the Filaments of the Nerves. The Debate having been managed a good while, by variety of Arguments on both sides, I (who had been used to suspect, that the greatest part of Disputes were more about the signification of Words, than a real difference in the Conception of Things) desired, That before they went any farther on this Dispute, they would first examine, and establish amongst them, what the Word *Liquor* signified. They at first were a little surprised at the Proposal; and had they been Persons less ingenuous, they might perhaps have taken it for a very frivolous or extravagant one: Since there was no one there, that thought not himself to understand very perfectly, what the Word *Liquor* stood for; which, I think too, none of the most perplexed names of Substances [the word, "liquor," I agree, is relatively well understood]. However, they were pleased to comply with my Motion; and upon Examination found, that the signification of that Word, was not so settled and certain, as they had all imagined; but that each of them made it a sign of a different complex *Idea*. This made them perceive, that the Main of their Dispute was about the signification of that Term; and that they differed very little in their Opinions, concerning some fluid and subtle Matter, passing through the Conduits of the Nerves; though it was not so easy to agree whether it was to be called *Liquor*, or no, a thing which when each considered, he thought it not worth the contending about.[110]

Here, Locke, at ease among his medical peers, recognized a universal problem in an arcane professional misunderstanding. Mediating in this company was straightforward enough. Just as easily, Locke must have hoped, other disputes more volatile could be settled among persons less "learned," less "ingenious"—and "less ingenuous."

But by what method? Boyle's physical chemistry could prove ultimate cause in the natural world—in theory, though not in practice.[111] And God's revelation could prove ultimate good in the moral world—also in theory and probably in practice as well. Perhaps, Locke speculated, men and women being "understanding, rational Beings," morality itself might be placed "*amongst the Sciences capable of Demonstration.*" Moral judgments "as incontestable as those in Mathematicks" might even be possible:[112] "*Where there is no Property, there is no Injustice,* is a Proposition as certain as any Demonstration in *Euclid*: For the *Idea* of *Property*, being a right to anything, and the *Idea* to which the Name *Injustice* is given, being the Invasion or Violation of that right; it is evident, that these *Ideas* being thus established, and these Names annexed to them, I can as certainly know this Proposition to be true, as that a Triangle has three Angles equal to two right ones."[113]

Several comments are indicated here.

First, as noted previously, Locke's system did show the marks of ancient and scholastic tradition in its assumption of ultimate cause in the natural world. The "historical, plain method" was clinical, not experimental; it was observational, not investigational; it could not produce every answer needed in a comprehensive treatment of human understanding. Rationalism—all thinking, no testing—still had a place.

Second, the "ultimate cause" Locke most favored—Boyle's corpuscular theory—turned out to be the forerunner of the best modern approximation to ultimate cause. So, Locke's choice may have been premature, but it was a good choice.

Third, Locke's use of the word "science" in reference to the making of moral decisions should remind us how ambiguous that term still was even among scientific revolutionaries and even in the writing of an archnominalist. The "science" of morality could not have been a Baconian science since it was strictly deductive. The "science" of morality was more like mathematics; it was more Cartesian. And, surely, moral argument was deductive, just as legal argument was deductive. But Locke had previously presented a Baconian analysis of religious variation, had described the effect of peer pressure in determining "the law of opinion or reputation," and had offered clinical observations on intellectual habituation.

Had Locke changed his mind about moral discourse by this point in his decades-long project? Probably not. "*Where there is no Property, there is no Injustice*" was a moral principle inducible from observation, as was its companion, "*No Government allows absolute Liberty.*"[114] Many of the difficulties adapting ethics to mathematical reasoning could "in a good measure be *remedied* by Definitions," though "other difficulties" might remain.[115] No doubt, they would. Locke was urged to expand on these notions, but he never did.[116] His commentary on the proper nature of moral investigation and moral judgment was never as full as it might have been.

Fourth, Locke's controversial emphasis on "property" appeared here as an emphasis on rights and on process.

The *Essay* ended with a coda of previous themes. Religious variation and even religious error were the faults of circumstance and teaching, not signs of immorality or unworthiness in the individual. Science had its limits, but experience must be valued over received knowledge. Scholastics would always stumble in an underbrush of syllogisms, while mankind's store of useful knowledge would be increased by tradesmen through craft, by mathematicians through logic, and by scientists through observation, experimentation, and appreciation of probability—such as the probability that "*Kin Kina*," or quinine, would save many lives if its "Virtue and right Use" were "made publick,"[117] as they were by Sydenham in England and by Locke in France.

Locke's collegial intentions aside, is there evidence that this difficult work ever found a reader, let alone a use, among physicians?

Yes, a little. The English physician and proto-utilitarian psycho-physiological philosopher David Hartley (1705–1757) acknowledged his debt to Locke's *Essay*, as well as to Newton's *Principia*, in the first chapter of *Observations on Man, His Frame, His Duty, and His Expectations* (1749).[118] The perpetually controversial French physician, *philosophe*, and satirist Julien Offray de La Mettrie (1709–1751) admired the *Essay* and welcomed its effect on his ideas.[119] Appreciations such as these may not have been exceptional within progressive Enlightenment medical communities. By the nineteenth century, when the philosophy of medicine began to be treated distinctly, empiricism was intellectually ascendant, though rationalistic therapeutic schools still teemed in the bushes. Bacon[120] and Newton[121] may more commonly have been cited as empiricism's major prophets than was Locke, but Locke's *Essay* was not unchampioned.

Sir Gilbert Blane, "First Physician to the King," cited Locke admiringly[122] in his 1825 treatise, *Elements of Medical Logick, or Philosophical Principles of the Practice of Physick.* The rejection of innate ideas was important for Blane and he defended Locke against the charge that such rejection meant "an end of all morality."[123] Blane was primarily concerned with "the sources of medical error," and he valued Locke's *Essay* especially—*especially*—for its section on nominalism:

> FIFTH SOURCE OF MEDICAL ERROR.
> THE AMBIGUITY OF LANGUAGE.
>
> *Danger of being guided by the Name instead of the Nature of a Disease, exemplified in Sea Scurvy, the Yellow Fever, and Dropsy.*
>
> As the end of language is the communication of thought, it is evident that there can be no such thing as correct reasoning, unless the same import be annexed to the same words, in the oral and written intercourse of mankind. A large proportion of all the false reasoning and controversy, which has existed among the learned and unlearned of all ages, has arisen from the want of a precise definition of words. The most valuable parts of the writings of Locke, are those which relate to the abuse of language. It is a subject upon which there is great room here to dilate; for none of the departments of practical knowledge have suffered more than medicine, from verbal disputations, engendered by ambiguous words and phrases. It falls under this

> head, therefore, to caution practitioners, particularly our younger brethren, against prescribing for the *name*, instead of prescribing for the *nature* of a disease; and it is proposed to illustrate this by a few examples, referring the reader to the chapters of Locke's Essay on the Understanding, which relate to this subject.[124]

Even the most tedious and "least medical" section of Locke's *Essay* could still find an audience among physicians—a few, at any rate, and presumably just in English—more than a century after Locke's death.

The scientific and clinical relevance of the *Essay* has been less apparent to most lay scholars. Throughout the book, references to scientific and clinical facts, theories, experiments, and procedures were prominent and pivotal; easily, they numbered in the scores. Yet generations of scholars have concluded—some through carelessness or systematic oversight but others[125] quite respectably through caution—that Locke's scientific orientation and clinical experience were incidental to the work.

We shall close this section by looking at a particular scholarly misjudgment found in Alexander Campbell Fraser's heavily annotated 1894 rendering of the *Essay*, a version still being marketed as late as the 1980s (well after release of the Nidditch edition) as "the only complete edition . . . currently in print." Locke was discussing the qualities and powers of bodies:

> The particular bulk, number, figure, and motion of the parts of fire or snow are really in them,—whether any one's senses perceive them or no: and therefore they may be called *real* qualities, because they really exist in those bodies. But light, heat, whiteness, or coldness, are no more really in them than sickness or pain is in manna. . . .
>
> A piece of manna of a sensible bulk is able to produce in us the idea of a round or square figure; and by being removed from one place to another, the idea of motion. . . . Besides, manna, by the bulk, figure, texture, and motion of its parts, has a power to produce the sensations of sickness, and sometimes of acute pains or gripings in us. That these ideas of sickness and pain are *not* in the manna, but effects of its operations on us, and are nowhere when we feel them not; this also everyone readily agrees to. And yet men are hardly to be brought to think that sweetness and whiteness are not really in manna; which are but the effects of the operations of manna . . . : as the pain and sickness caused by manna are confessedly nothing but the effects of its operations on the stomach and guts.[126]

What was "manna"? Fraser may not have known. In four footnotes to the "manna" paragraphs, Fraser cited George Berkeley (1685–1753) and David Hume (1711–1776) to no advantage and then declared, in reference to "a power to produce the sensations of sickness," that "[i]t signifies that the continuously active Divine Reason, immanent in things, is about to produce those 'sensations' in us."[127]

"Divine Reason"? As in "manna from heaven"? Might not such symbolism, such an elaborate conceit, have been thought odd in a book exhibiting an "historical, plain method," a book in part devoted to the precise, nonrhetorical use of words?

Had he recalled the classical medical literature, Fraser might have found Locke's discussion of manna reminiscent of a Hippocratic author's discussion of cheese in the treatise *On Ancient Medicine*;[128] manna was not cheese, of course, but neither was cheese "Divine Reason." More directly, Fraser could have had access to many of Locke's medical papers but, in fairness, not to the mass of his medical journals. If he had managed a look, he might have noticed manna being compounded,[129] concocted,[130] and otherwise prescribed or commended[131] by Locke in the course of his clinical practice. Yet even without clues from Locke's patient progress notes, Fraser could still have sought "manna" in a dictionary. Samuel Johnson described it correctly and at length.[132] Manna was, and is, a sweet crystalline exudate from the Eurasian ash tree. It is now known to contain mannitol, a hexahydroxy alcohol named after its source. Manna, ingested in sufficient quantity, obligates by osmotic pressure enough free water in the lumen of the gut to stimulate colonic evacuation. Manna was not "Divine Reason." Manna was a laxative.

Was the empirical epistemology of *An Essay concerning Human Understanding* affected by Locke's scientific orientation and clinical experience? Yes, it was. Did the empiricism of the *Essay* appeal particularly to later life scientists and clinicians? Yes, it did.

Empiricism and Locke's Liberal Political Ethics

While in exile in Holland, Locke attended the regular meetings of a group of Remonstrant physicians. Through this group, Locke came to attend the dissection of a lioness, killed by cold at the Amsterdam zoo during the remarkably severe winter of 1683–1684. Also attending the dissection was an amateur scientist of whom Locke had already heard, a famous Remonstrant theologian, Philip van Limborch,[133] soon to become one of Locke's most valued friends. By the winter of 1685–1686, Locke, implicated in the Monmouth Rebellion and listed for extradition, had taken refuge in a fellow physician's house and had assumed a false name. But Locke—or "Dr. van der Linden"—and Limborch continued to meet.[134] At Limborch's urging, or, at any rate, during a period of close association with Limborch, Locke set aside work on the *Essay* long enough to write *Epistola de Tolerantia,* his first *Letter on Toleration.* The *Letter* came between the completion of Book Two and the editing of Book Three[135] and, therefore, between Locke's observations on the "disease" called "prejudice" and his argument "that the main of . . . dispute[s] was about the signification of . . . term[s]," many of which were just "not worth the contending about." The *Letter* was written secretly and, in Holland in the spring of 1689, published anonymously.[136]

Adoption of an empirical world view has never been prerequisite to the development of a tolerant religious attitude. A wholly nonempirical provenance could be written for the *Letter.* But such a provenance could never be complete. In the *Letter* were heard many echoes of the *Essay.* Humans were rational animals with spontaneous thoughts. Their thoughts could be erroneous, but their thoughts were always only their own. Thoughts were not "innate" and could never be imposed against the will, but their specific content, as anyone could observe, was largely a function of circumstance and teaching; Turks were always going to be Moslems. Since it was not reasonable to think that "men would owe their eternal happiness or misery to the places of their nativity,"[137] the specific content of religious thoughts could not be nearly as important as intolerant people claimed. Even if determined by misfortune, religious conviction could still be only a matter of conscience. Logically, then, religious practice should be self-regulated in freely congregated and conscientiously homogeneous churches and communities, and the civil state should have nothing to say on the subject—except, of course, in the case of Catholics, acting, as they were wont to do, as agents of a foreign prince, the bishop of Rome.

Any systematic observer of human behavior could see the error in religious oppression:

> I know that seditions are frequently raised upon pretence of religion, but it is true that for religion subjects are frequently ill treated, and live miserably. Believe me, the stirs that are made proceed not from any particular temper of this or that church or religious society, but from the common disposition of all mankind, who when they groan under heavy burthen endeavor naturally to shake off the yoke that galls their necks. Suppose this business of religion were let alone, and that there were some other distinction made between men and men upon account of their different complexions, shapes, and features, so that those who have black hair (for example) or grey eyes should not enjoy the same privileges as other citizens; that they should not be permitted either to buy or sell, or live by their callings; that parents should not have the government and education of their own children; that they should either be excluded from the benefit of the laws, or meet with partial judges; can it be doubted but these persons, thus distin-

guished from others by the color of their hair and eyes, and united together by one common persecution, would be as dangerous to the magistrate as any others that had associated themselves merely upon the account of religion? Some enter into company for trade and profit, others for want of business have their clubs for claret. Neighbourhood joins some, and religion others. But there is only one thing which gathers people into seditious commotions, and that is oppression.[138]

Locke's most famous political work, *Two Treatises of Government*, was, like the *Letter on Toleration*, written secretly and published anonymously. The appearance of *Two Treatises* in 1689 seemed conveniently to justify the Glorious Revolution, but, as Peter Laslett has argued convincingly, the book must have been finished, save for a few flourishes, long before 1688.[139] Work on the *Essay* and work on the *Two Treatises* must have proceeded alternatively, if not simultaneously. For part of its prepublication history, *Two Treatises* was probably hidden under the false title, *De Morbo Gallico*, the "French disease," meaning, of course, syphilis.[140] Locke did not acknowledge authorship until 1704, the year he died, and then only indirectly.[141] *Two Treatises* was a politically dangerous book in Locke's day, and it has remained an academically controversial one in our own.

The *First Treatise* argued against Sir Robert Filmer's then-prominent scriptural defense of absolute monarchy, a defense that Filmer had fashioned cleverly: "Like a wary Physician, when he would have his Patient swallow some harsh or *Corrosive Liquor*, he mingles it with a large quantity of that, which may dilute it; that the scatter'd Parts may go down with less feeling, and cause less Aversion."[142] As portrayed by Locke, "Sir *R. F.*'s great Position is, that *Men are not naturally free.*"[143] Filmer saw God giving all gifts to fathers, husbands, and monarchs, and only then through their graciousness to children, wives, and subjects, none of whom were bearers of rights in the least degree.

Locke saw nothing of the kind. God's gifts were birthrights given to all individually and in common. All children were born free and equal, and freedom and rough equality were meant to persist throughout their lives. They even had a responsibility as adults to redistribute their fortunes within their communities, lest bad luck too severely unbalance natural equality:

> But we know God hath not left one Man so to the Mercy of another, that he may starve him if he please: God the Lord and Father of all, has given no one of his Children such a Property, in his peculiar Portion of the things of this World, but that he has given his needy Brother a Right to the Surplusage of his Goods; so that it cannot justly be denyed him, when his pressing Wants call for it. And therefore no Man could ever have a just Power over the Life of another, by Right of property in Land or Possessions; since

'twould always be a Sin in any Man of Estate, to let his Brother perish for want of affording him Relief out of his Plenty. As *Justice* gives every Man a Title to the product of his honest Industry, and the fair Acquisitions of his Ancestors descended to him; so *Charity* gives every Man a Title to so much out of another's Plenty, as will keep him from extream want, where he has no means to subsist otherwise; and a Man can no more justly make use of another's necessity, to force him to become his Vassal, by with-holding that Relief, God requires him to afford to the wants of his Brother, than he that has more strength can seize upon a weaker, master him to his Obedience, and with a Dagger at his Throat offer him Death or Slavery.[144]

Husbands could hold no more than a "Conjugal Power" over their wives, not a "Political" power.[145] And fathers could not "by begetting them, come by an Absolute Power over their Children."[146] Fathers were not, as Filmer claimed, true givers of life to their sons and daughters: "How can he be thought to give Life to another, that knows not wherein his own Life consists?"[147] This rhetorical question was answered with assertions Sydenham must have made to Locke again and again during smallpox rounds, assertions which would have been directed professionally by Sydenham against Willis, politically by Locke against Filmer: "Philosophers are at a loss about it [the essence of life, physiological and procreative] after their most diligent enquiries; And Anatomists, after their whole Lives and Studies spent in Dissections, and diligent examining the Bodies of Men, confess their Ignorance in the Structure and Use of many parts of Mans Body, and in that Operation wherein Life consists in the whole."[148]

It was "hard to imagine the rational Soul should presently Inhabit the yet unformed Embrio, as soon as the Father has done his part in the Act of Generation." And it was just as hard to imagine denying the mother "an equal share in the begetting of the Child."[149] Abuses of children were not to be seen as evidence of parental power over their offspring but as unnatural crimes:

The Dens of Lions and Nurseries of Wolves know no such Cruelty as this: These Savage Inhabitants of the Desert obey God and Nature in being tender and careful of their Off-spring: They will Hunt, Watch, Fight, and almost Starve for the Preservation of their Young, never part with them, never forsake them till they are able to shift for themselves; And is it the Priviledge of Man alone to act more contrary to Nature than the Wild and most Untamed part of the Creation? . . . [God] has in all the parts of the Creation taken a peculiar care to propagate and continue the several Species of Creatures, and makes the Individuals act so strongly to this end, that they sometimes neglect their own private good for it, and seem to forget that general Rule which Nature teaches all things of self Preservation, and the Preserva-

> tion of their Young, as the strongest Principle in them over rules the Constitution of their particular Natures. Thus we see when their Young stand in need of it, the timorous become Valiant, the Fierce and Savage Kind, and the Ravenous Tender and Liberal.[150]

Divine law and natural law were entirely compatible. Indeed, it was the observation of nature that best revealed God's intention. The surest way to follow divine law in government was to follow nature in civil law. It was not enough to follow custom, for custom, just as in the *Essay*, was a repository of ethical error:

> And when Fashion hath once Established, what Folly or craft began, Custom makes it Sacred, and 'twill be thought impudence or madness, to contradict or question it. He that will impartially survey the Nations of the World, will find so much of the Governments, Religions, and Manners brought in and continued amongst them by these means, that he will have but little Reverence for the Practices which are in use and credit amongst Men, and will have Reason to think, that the Woods and Forests, where the irrational untaught Inhabitants keep right by following Nature, are fitter to give us Rules, than Cities and Palaces, where those that call themselves Civil and Rational, go out of their way, by the Authority of Example.[151]

The natural freedom and equality of men (and women) notwithstanding, there did exist an obligation to obey proper authority. This being the case, clarity in the constitution of authority was necessary. Otherwise, "Men too might as often and as innocently change their Governours, as they do their Physicians, if the Person cannot be known, who has a right to direct me, and whose Prescriptions I am bound to follow."[152] The details of this obligation, the concept of its reciprocity, and the conditions under which it could "innocently" be ignored would all be presented in the *Second Treatise.*

But one item still detains us in the *First Treatise.*

Reluctance to accept a link between Locke's empiricism and his political ethics is strong among those who see as inconsistent the rejection of innate ideas in the *Essay* and the description of natural law in the *Two Treatises.*[153] Locke was not being inconsistent; both concepts appeared in both works.

In the *Essay,* Locke described "natural tendencies" distinct from "innate ideas," by which latter term he meant specific axioms inborn in rational minds. In the *Two Treatises,* "natural tendencies" appeared as the desires "God Planted in Men, and wrought into the very Principles of their Nature." The desires for "Copulation"[154] and "Self-preservation"[155] were among the most obvious. Implanted desires were not at all the same as "innate ideas," a term with which Locke's name was soon to become famously associated in the *Essay*, a term he may not have wanted to use in the anonymously authored *Two Treatises.* Recalling the *Essay*, Locke noted that, since the privacy of filial inheritance was respected not just by "common tacit Consent" but by universal "Common Practice," parental provision for children surely had a "Natural" cause; and, yet, parental provision for children had been made a statutory obligation, as was "evident from the Law of the Land."[156] The existence of a legislative mechanism for compulsion was proof that a "natural tendency" was not an "innate idea." Filmer's version of the "innate ideas" theory was criticized just as it would have been in the *Essay*, and no less pointedly: "Strange! that *Fatherly Authority* should be the only Original of Government amongst Men, and yet all Mankind not know it."[157]

Far from being an inconsistency between the *Essay* and the *Two Treatises*, the concept of natural tendencies or implanted desires was one of the strongest links. It was from the empirical observation of implanted desires that natural law was induced: "God Planted in Men a strong desire . . . of propagating their Kind, and continuing themselves in their Posterity, and this gives Children a Title, to share in the *Property* of their Parents, and a Right to Inherit their Possessions."[158] God had created and had implanted natural law, and people could know it through observation and induction, and they were obliged to obey it:

> The *State of Nature* has a Law of Nature to govern it, which obliges every one: And Reason, which is that Law, teaches all Mankind, who will but consult it, that being all equal and independent, no one ought to harm another in his Life, Health, Liberty, or Possessions. . . . And being furnished with like Faculties, sharing all in one Community of Nature, there cannot be supposed any such *Subordination* among us, that may Authorize us to destroy one another, as if we were made for one anothers uses, as the inferior ranks of Creatures are for ours. Every one as he is *bound to preserve himself*, and not to quit his Station wilfully; so by the like reason when his own Preservation comes not in competition, ought he, as much as he can, *to preserve the rest of Mankind*, and may not unless it be to do Justice on an Offender, take away, or impair the life, or what tends to the Preservation of the Life, Liberty, Health, Limb or Goods of another.[159]

Here were many fresh ideas. Natural law and reason were one and the same, and, so, natural law was self-revealing to rational men and women. Self-evidently, humans should not harm one another, nor should one use another as one might use a lower animal—a "nonsubordination" rule echoing Celsus on research ethics and anticipating Immanuel Kant (1724–1804) on means and ends.

Punishment could serve a deterrent intention, Locke said, but punishment of a transgressor, even in the State of Nature, still had to correspond to the severity of the transgression punished.[160] Deterrence could not be disproportional to risk.

Locke agreed with Hobbes that "all *Princes* and Rulers of *Independent* Governments all through the World, are in a State of Nature."[161] But Locke did not thereby exclude the possibility of limited compacts being reached among men and women who lacked a common judge. "Cooperation under anarchy," as it might now be called, was a possibility, "For truth and keeping of Faith belongs to Men, as Men, and not as Members of Society."[162] More fundamentally, Locke did not accept Hobbes's equation of the states of nature and war:

> And here we have the plain *difference between the State of Nature, and the State of War*, which however some men have confounded, are as far distant, as a State of Peace, Good Will, Mutual Assistance, and Preservation, and a State of Enmity, Malice, Violence, and Mutual Destruction are one from another. Men living together according to reason, without a common Superior on Earth, with Authority to judge between them, is *properly the State of Nature*. But force, or a declared design of force upon the Person of another, where there is no common Superior on Earth to appeal to for relief, *is the State of War:* And 'tis the want of such an appeal gives a Man the Right of War even against an *aggressor*, though he be in Society and a fellow Subject. Thus a *Thief*, whom I cannot harm but by appeal to the Law, . . . I may kill . . . because the Law . . . permits me my own Defence, [as does] the Right of War, because the aggressor allows not time to appeal to our common Judge, nor the decision of the Law. . . . *Want of a common Judge with Authority, puts all Men in a State of Nature: Force without Right, upon a Man's Person, makes a State of War*, both where there is, and is not, a common Judge. . . .
>
> To avoid this State of War . . . is one great *reason of Mens putting themselves into Society*, and quitting the State of Nature.[163]

But men and women could not avoid a state of war by putting themselves into a society ruled by an absolutist government. Here Locke's conception most radically departed from Hobbes's. Absolutism did not avoid a state of nature. Absolutism *was* a state of nature and was "*inconsistent with Civil Society*, and so can be no Form of Civil Government at all."[164]

Locke thought it "very clear, that God . . . *has given the Earth to the Children of Men*, given it to Mankind in common," not to monarchs, not to the fortunate, not to an elect. But how then could humankind have converted a common gift into multiple private properties justly and "without any express Compact of all the Commoners"?[165] Locke's answer was his famous "labor theory of value," a concept foreshadowed by a multifactorial labor-and-land theory of value found in *Natural and Politi-*

cal Observations upon the Bills of Mortality, published in 1662 under the name of John Graunt but widely attributed then and since to Petty,[166] whose political philosophy and political advice were, despite a long friendship with Boyle, implicitly Hobbesian: mechanistic, state-centered, metaphorically corporate, functionally amoral—and notably unconcerned with the rights of individuals, especially those individuals who were poor or hungry or Irish.[167] Still, Locke admired Petty's demography and empirical economics. He had a 1667 copy of the *Observations* as well as some of Petty's "political arithmetick" tracts; while in France, he secured for his own study the bills of mortality for Paris, and, while in Rotterdam, he kept his own weekly burial records.[168]

In Locke's new theory, all people owned at least their own bodies, therefore their own labor, and, in turn, the product of their labor up to and including as much as could be used "before it spoils."[169] Thus, the "Industrious and Rational" profited within the bounds of natural law. More lately, alas, the use of money had made it possible for people to own far more than what they could themselves use, far more than what they had a natural right to own. This change had been an unfortunate one, for, previous to it, private ownership must almost always have improved the general welfare, at least in regions of surplus resources.[170]

If all people owned their bodies, did they have an arbitrary power over their own lives? An affirmative answer would have been consistent with some of Locke's philosophy and with much of its subsequent libertarian interpretation. A negative answer could have suggested a quasi-communitarian motive, a Christian motive, or even a Hippocratic motive, the last recalling the ancient parentalistic injunction against assisting or encouraging a patient's suicide. From comments elsewhere in the *Second Treatise*, we can say only that Locke was no libertarian: "[N]o Body has an absolute Arbitrary Power over himself . . . to destroy his own Life."[171]

Ownership of body was one thing, but inheritance of property another. The right to bequeath real property, the right to secure one's patrimony against the confiscatory power of king, commonwealth, or conqueror, depended upon the reality of natural law, depended upon the *fact* of implanted desire. The "tenderness for their Off-spring" that "God hath woven into the Principles of Humane Nature"[172] had to be demonstrable. And it was:

> This Rule, which the infinite wise Maker hath set to the Works of his hands, we find the inferiour Creatures steadily obey. In those viviparous Animals which feed on Grass, the *conjunction between Male and Female* lasts no longer than the very Act of Copulation: because the Teat of the Dam being sufficient to nourish the Young, till it be able to feed on Grass, the Male only

begets, but concerns not himself for the Female or Young, to whose Sustenance he can contribute nothing. But in Beasts of Prey the *conjunction* lasts longer. . . . The same is to be observed in all Birds (except some domestick ones . . .). . . .

And herein I think lies the chief, if not the only reason, *why the Male and Female in Mankind are tyed to a longer conjunction* than other creatures, *viz.* because the Female is capable of conceiving, and *de facto* is commonly with Child again, and Brings forth too a new Birth long before the former is out of a dependancy for support on his Parents help. . . . Wherein one cannot but admire the Wisdom of the great Creatour.[173]

Overlooking Locke's pre-Darwinian teleological assumptions, the operative naturally selective pressures induced in this passage have been induced correctly, or almost correctly. If there is a "natural law theory" consistent with modern science, then it must be consistent first with evolutionary biology. Locke's argument, again with the cited reservation, would meet this condition fairly well.

Locke's version of natural law was a codification of human ethical intuition. Its principal codes were induced from observation of *Homo sapiens* as one live-born species among many species, some quite similar, some wildly different. Though drawn from the wisdom of the forest and cottage—and sometimes the barnyard—its implications were protean, extending even to civil and international disturbance.

The person who recognized the universality of the Law of Nature would also recognize that "he and all the rest of *Mankind are one Community*. . . . And were it not for the corruption, and vitiousness of degenerate Men, there would be no need of any other; no necessity that Men should separate from this great and natural Community, and by positive agreements combine into smaller and divided associations."[174]

"Force without Right" had transformed a universal community into a community of nations. And it could also temporarily transform a state of nature into a state of war, even, under certain circumstances, within societies. In disputes between societies—in wars—the prerogatives of conquest depended entirely upon the just cause of the conqueror, "From whence 'tis plain, that he that *Conquers in an unjust War, can* thereby *have no Title to the Subjection and Obedience of the Conquered.*"[175] War guilt could be ascribed to overthrown leaders but probably not to their subordinates generally.[176] The rights of the just conqueror were "despotical" regarding the lives of defeated combatants, but not regarding their property or their families; the conqueror was even obliged to subtract from the fair reparation of his own losses enough to maintain the children of a vanquished aggressor. And conquest, even just conquest, could not transfer dominion.[177]

We should note that the despotical power of conqueror over defeated combatant justified slavery. Locke was an investor in a certain Bahamian enterprise,[178] and he seems to have held the view that Africans enslaved to serve the Caribbean trade had forfeited their natural freedom by engaging unsuccessfully in aggressive warfare. Locke saw justifiable slavery as an alternative to justifiable execution. He did not approve of hereditary slavery,[179] though that became the pattern of servitude in the economy in which he invested.

One of Locke's central observations was the nearly universal emphasis placed on initial and continuing consent in successful, voluntary human societies and associations. Power unavoidably implied fiduciary responsibility, and a violation of faith always justified withdrawal of consent. So it was with monarchs no differently than with officers of a commonwealth, with "Kings" no differently than with "Constables," since "exceeding the Bounds of Authority is . . . so much the worse in him, in that he has more trust put in him, has already a much greater share than the rest of his brethren, and is supposed from the advantages of Education, imployment and Counsellors to be more knowing in the measures of right or wrong."[180] So it was to be also, one suspects, with physicians.

Some scholars have denied a link between the empiricism of the *Essay* and the natural-law rationalism of the *Two Treatises.* Laslett has suggested that Locke refused to admit authorship of the *Two Treatises* well beyond the years of personal danger to avoid being embarrassed by inconsistencies with the *Essay.*[181] This is a serious issue. If his empiricism and his natural law did mutually exclude one another, Locke was certainly a confused philosopher, perhaps even a dishonest man, and his standing as a contributor to the political-ethical tradition of the life sciences would have to be disallowed.

The *Essay* and the *Two Treatises* were certainly different projects, but they were not seriously inconsistent. Combined, they made something of a whole—which is not to suggest that their combination was ever intended, only that its accomplishment would not have been a wrenching task.

In the *Essay,* Locke had set out "not to know all things, but those which concern our Conduct,"[182] and he had described three moral standards by which conduct could be judged. The third standard, "the law of opinion or reputation," he explored in the *Essay* itself; it was a standard describable only by induction from observed practice, and its discussion fit well in an empirical work. The first standard, divine law, he touched on only lightly in the *Essay*; the second standard, civil law, he ignored, except to the extent that civil law was a function of "the law of opinion or reputation." It was in the *First Treatise* that divine law would have its closest examination, and in the *Second Treatise* that civil law would be explored.

The first and second moral standards were traditional ethical topics, and they had to be discussed deductively, not discussed solely from observation. Even in the *Essay*, a book exalting *inductive* thought, Locke had asserted that moral argument was *deductive.* He had even predicted that the reliability of deductive moral argumentation would someday match the reliability of mathematics.

Locke deduced selected requirements of divine law from their only legitimate source, the revealed word of God. An attack on Filmer's book provided the framework. But from what starting place did Locke deduce the requirements of civil law? God's revelations were less serviceable here—and more dangerous. If church and state were to be separate, if religious toleration was to thrive, then Locke needed a different set of first principles from which to deduce civil law. Natural law provided them.

God's law was easily made to contain Nature's law. And the principles of natural law could be induced from the observation of nature itself, induced from the observation of *Homo sapiens*, induced from the observation of human understanding. "Is the Law of nature Knowable by the light of nature? It is."[183] "Can Reason arrive at a knowledge of the law of nature through sense experience? It can."[184] So asked and answered Locke secretly in *Questions concerning the Law of Nature*, a work till recently obscure.[185]

It was still the human being of the *Essay*—rational, teachable, admirable, and corruptible—that occupied the *Two Treatises.* It was to him that divine law applied. It was from his qualities and habits, and from those of his cousin species, that the principles of natural law were induced, "sociobiologically." And it was for the protection of his "Property"—his life and what tended to the preservation of his life, his liberty, his health, his limb, his labor, his land, his possessions, his goods, and, collectively, his prerogative and his dignity—that civil law was *deduced* from the *induced* principles of natural law.

The induction of hypotheses was routine in Baconian science, and several of the hypotheses induced in the seventeenth century quickly became generalized as "laws." Deriving civil laws from the induced laws of nature was similarly unexceptional. Hobbes, in a single, well-organized book, *Leviathan*, had presented a striking vision of human nature and had then derived from it a system of political ethics. Locke did the same thing, but in an essay, a letter, and two treatises, for the most part poorly organized and edited, and published in no particular order.

Hobbes far exceeded Locke as a conductor of large literary forces, yet he ended up drawing a political-ethical conclusion repugnant even to most admirers: the proper preventive for civil discord was authoritarianism. Locke's preventives for civil discord—clear thinking, tolerance, respect for human rights—underlay a political ethics that was to prove enormously more helpful.

Why did the lesser craftsman prove the greater builder?

Let us assume that both were honest workmen, that both conscientiously deduced political ethics from theories of human nature. Hobbes, self-assessment aside, was out-of-date in the sciences generally and amateurish in the life sciences specifically. He started from a pseudo-physiological theory of human nature hard to describe now except as ridiculous. Locke, a life scientist and clinician by education, training, practice, self-image, and contemporary reputation, started from a far more reasonable theory of human nature, a theory with which advancing scientific understanding would remain broadly sympathetic.

However, in political philosophy, the honest-workman assumption is a strong assumption indeed. We would more safely figure that Hobbes and Locke both knew all along where their derivations would lead. The quality of their respective endpoints might then be judged by the starting-point distortions each "conclusion" required for its contrivance. The starting-point distortions required by Hobbes were severe, those required by Locke comparatively mild.

Either way, the advantage went to Locke.

For his originality, and for the sloppiness that sometimes attended it, Locke has paid the standard posthumous critical price. He has been vilified for a range of contradictions, insincerities, errors, and evil legacies. Some have complained that he did not faithfully or gratefully transmit the natural-law tradition that he and his contemporaries had received from theologians and philosophers, such as Aquinas, and, particularly, that he owed less to one, Richard Hooker (1554–1600), and more to another, Hobbes, than he claimed or admitted.[186] Had they not been refutable[187] or, at any rate, reducible[188] on biographical and textual grounds, the Hooker-and-Hobbes charges, which came to prominence in the bitterest stretch of the Cold War, would have made Locke less distinct from Hobbes and, so, less an optimist, less a liberal, and, for some purposes, little more than a minor political realist. On matters of character, it has been argued, but not surprisingly, that Locke said less than he meant (and said the remainder less than candidly) when fearing for his freedom, which he managed never fully to lose. It has also been said—it *is* said—that Locke inspired or preexcused hedonism and power-lust and greed in generations of rascals not otherwise prone to self-indulgence or political dominance or contractual sharp dealing.[189] Though Locke's name has often been invoked to bless—and his influence has been accused of explaining—sophisticated selfishness, his philosophy, read anew, and read with an eye for scientific process and clinical experience, makes sense. And it makes peace.

Did Locke's empiricism affect the "liberal" political ethics of the *Letter on Toleration* and the *Two Treatises of Government*? Yes. Then, was Locke's liberalism the inevitable product of his physicianhood? His political-ethical views had been Hobbesian as late as 1660, the year he began working with Boyle, and his scientific and clinical associations did affect his thinking fundamentally. But "inevitable product"? No. It is enough to say that Locke's physicianhood and Locke's liberalism bound to each other with high affinity.

Locke and Liberalism in the Life Sciences

It is not true, as some have supposed, that Locke's medical accomplishments were wholly overlooked till the modern era. Scattered references to his clinical work occurred in the eighteenth century, and, in 1829, a new biography and a monograph both noted the high regard Sydenham had had for Locke as a physician and disputed the notion that Locke had been medically inactive.[190]

Did the 1829 revisions in the lay literature correct an equivalent amnesia in corporate medical knowledge or in the medical literature? Romanell has suggested that they did and that, prior to 1829, physicians had been as likely as laymen to disregard Locke's medical past. Romanell has cited as evidence a note in the *Lancet*. The issue of June 20, 1829, recorded a presentation by Locke's latest biographer, Lord King, a distantly related peer:

> THE CELEBRATED LOCKE AS A PHYSICIAN.
>
> At the conversazione held in the College of Physicians on Monday, June 1st, those amongst this "learned body" who were ignorant of the fact, had an opportunity of hearing it confirmed beyond all doubt, that the great John Locke belonged to the profession, Lord King (who was present) having put into the hands of Dr. Clarke, of Saville Row, an old French almanac, formerly the property of Locke, in which there were blank pages interleaved, containing notes and memoranda, settling all question on the subject. The notes . . . contained [Locke's account of a case] in all respects resembling *tic douloureux*.[191]

Romanell's citation of this entry may inadvertently have accentuated the novelty of Lord King's gift, "'learned body'" having been copied as "learned body"—without the *Lancet*'s own quotation marks.[192] The tone of the original is not revelatory but waggish: how could a "body" including members so ignorant be called "learned"? And, strictly, this item from 1829 is not the earliest reference to Locke in the medical literature extant. Leafing through older numbers of the same journal reveals others. Within a year of the *Lancet*'s 1823 introduction, Locke's name appeared in the transcript of the Hunterian Oration to the Royal College of Surgeons:

> Gentlemen,—This anniversary Oration was founded to commemorate the birth of John Hunter [English anatomist and surgeon, 1728–1793], a man whose superior mental powers have contributed more to elevate the profession of surgery than those of any other individual. Such, indeed, was the superiority of his talents, that he is justly entitled to be distinguished as a man of genius. This is a title which should be applied only to men of transcendant abilities, on whom mankind in general have agreed to confer it; such men, for instance, as Locke, as Newton, or Harvey, of the superiority of whose powers there can be no dispute.[193]

Was Locke mentioned here, and mentioned first, because the speaker counted him a man of science and medicine? Or philosophy and politics? We do not know. A letter to the editor three weeks later commented on this address, disputing Hunter's credentials as a genius[194] but making no mention of Locke, Newton, or Harvey.

There is, though, another reference to Locke antedating the 1829 entry, and this one suggests—as, indeed, the 1829 entry suggests—that physicians of the day, probably including the Hunterian Orator, were aware of Locke's membership in their profession and were proud of it. Proud enough to find diverting a report of Lockeana soon to become available: "A literary Gentleman of our acquaintance has recently, by the death of a relation, come into possession of numerous unpublished manuscripts, and the minor correspondence of that celebrated physician, and, we may add, still more celebrated *metaphysician*, JOHN LOCKE. It is his intention to prepare a collection of these manuscripts for the press as soon as possible."[195]

Among a medical readership, Locke was both a "celebrated physician" and a "still more celebrated *metaphysician*" sometime *before* 1829. That said, the events of 1829 did shed indirect light on the wider problem of Locke's medical obscurity among a lay readership. An item from the following spring reflected this phenomenon.

It happened that Lord King's donation of the old French almanac seeded a curiosity about Locke-the-physician-philosopher in at least one spry antiquarian:

> LOCKE AS A PHYSICIAN
>
> *To the Editor of* THE LANCET.
>
> Sir,—In the early part of last June, a paper was read at the College of Physicians, containing an account of a medical case, treated by John Locke, a man, in my opinion, second only to Lord Bacon. The subject caught my

attention at the time, and as I am in the frequent habit of visiting the library of the British Museum, I looked over Ayscough's Catalogue, to see if other cases might not be found. The search was a gratifying one, and the various extracts I made, have been allowed to remain in my commonplace book, thinking that Lord King, in his life of Locke, would publish, if not the self-same extracts, at least others bearing on the same point. This, however, has not been done; if, therefore, you think them worthy of a place in THE LANCET, you will oblige me by printing them.

Your obedient servant,

JOHN P----E.

London, April 10th, 1830.[196]

There followed ten "extracts" of correspondence consistent with the notion that Locke had actually been engaged in the practice of medicine.[197] But more helpful than John P----e's trip to the library was his willingness to suggest in print that Lord King was neglecting the further documentation of Locke's medical identity. Lord King's example would be repeated again and again by lay scholars. Even H. R. Fox Bourne, whose two-volume 1876 biography[198] confidently recalled to scholarly attention Locke's medical career, did not pursue matters avidly.

But two physicians did. Edward T. Withington and William Osler, excited by Fox Bourne's book, leapt at the opportunity publicly to claim, or to *re*claim, for their profession an illustrious "lost" son.

Withington admired Locke as "the most celebrated man that ever belonged to the British medical profession," surpassing "Harvey, Sydenham, Hunter or [Edward] Jenner [(1749–1823), smallpox vaccination pioneer]."[199] Withington, co-editor and co-translator of a standard version of the Hippocratic opus, saw Locke the clinical historian as a greater Hippocratic than Sydenham himself.[200]

In the case of Osler, fuller appreciation of Locke's physicianhood had deep personal significance and, Osler being Osler, far-reaching consequences for the profession.

As Locke had been the great synthesizing figure of the Enlightenment, so Osler was the great synthesizing figure of modern medicine. He must have been the best informed part-time medical historian of his day. He was without dispute the most accomplished clinician and the most quoted professor of medicine in the English-speaking world. The new specialty of "internal medicine" was in part his conception, and he authored its most authoritative nineteenth-century text, a book still numbered proudly as the first edition of a leading late-twentieth-century reference. He led the first modern academic "internal medicine" department in the first modern medical school, "The" Johns Hopkins. He helped establish pediatrics as an academic discipline. He elevated to a still-intimidating height the "bench-to-bedside" standard of clinical practice and clinical research, and he deluged the clinical literature with original contributions. Most significantly, he popularized the "Oslerian system" of medical and surgical housestaff training and taught personally or otherwise affected a remarkably high proportion of the twentieth century's most influential medical researchers, practitioners, and educators.

Osler's career was so valued and his loss to post-influenza pneumonia, empyema, and pulmonary abscess[201] at the end of 1919 so widely regretted that Harvey Cushing—fifty years old when his mentor died, Surgeon-in-Chief at the Peter Bent Brigham Hospital, Boston, pioneer neuroendocrinologist, and, at the time, the world's most successful brain surgeon—stayed out of the operating room and the laboratory long enough to write the man's biography: 1,413 pages, the winner of a 1926 Pulitzer Prize and still a favorite presentation gift within the medical community.

On January 16, 1900, William Osler, M.D., Fellow of the Royal Society and Professor of Medicine at The Johns Hopkins University, addressed the Students' Societies of the Medical Department of the University of Pennsylvania. His treatise, "John Locke as a Physician," was based on his own meticulous reading of manuscripts in the British Museum and Public Record Office, not just on secondary sources. Osler opened by citing Locke's most widely appreciated contributions: "common sense" for "the working-day world" and "the Epistle on Toleration." He commented on Locke's career as an active practitioner, consultant, medical essayist, and partner to the great Sydenham. He reviewed in detail what he judged to have been Locke's outstanding management of difficult cases, especially the case of Lord Ashley's acutely suppurative hepatic hydatid cyst. "Perhaps the greatest . . . English philosopher," concluded Osler, "we may claim Dr. Locke as a bright ornament of our profession, not so much for what he did in it, as for the methods he inculcated and the influence which he exercised upon the English Hippocrates [Sydenham]."[202]

Osler's interest in Locke went well beyond this occasion, as allusions in the medical literature[203] and as references throughout Cushing's biography attest.

Later in 1900, the *Lancet* printed "John Locke as a Physician" in full and followed the next week with editorial comment:

> LOCKE'S work in philosophy and in the vindication of liberty has been so important that we have been too apt to overlook and to forget his connexion with our profession. It has been left for a physician living more than two centuries after him, and himself at the very front of modern medicine, to remind us of that connexion forcibly. . . . Dr. OSLER has now shown that in Locke English medicine had potentially another SYDENHAM. . . . And

> we are proud to think that the world is much indebted to his medical knowledge—or, if we may so say, to his medical mind—for the common sense which characterizes his philosophy and which is apparent in every page of his greatest work.
>
> In the few remarks we make [we emphasize] our gratitude to Dr. OSLER and our own pride in the fact that his hero really and historically belongs to the medical profession.[204]

The sentence beginning "And we are proud to think that the world is much indebted" bears rereading. It identifies the "medical mind" with "common sense" in philosophy. Somewhat too ingenuously, perhaps, as was the manner among self-assured late Victorians. But also in a way recalling the relationship between physicianhood and Lockean liberalism displayed in Locke himself: not a causative sequence so much as a strong natural attraction—"high-affinity binding," as we have called it.

Did "liberalism" enter the political-ethical tradition of the life sciences after Locke? Yes, it did. Did that liberalism persist into the "modern" era? Yes, though it faded between Locke and Osler. Did life scientists and physicians ever associate liberalism with Locke? Yes, though fitfully.

V

FROM NONLIBERAL ALTERNATIVES TO THE LIBERAL REESTABLISHMENT

IF A POLITICAL ethics of the life sciences had been written *in English* in the first years of the twentieth century, Lockean liberalism, adapted to bear "the white man's burden," would have dominated many arguments. But not all. Moreover, if a contemporaneous political ethics had been written *in German*, Lockean liberalism might have been missing altogether. To an uncomfortable degree, the reaffirmation of Lockean liberalism that took place within the next half century was a spoil of war. But it was also a triumph of good science over bad, of objectivity over prejudice.

Let us take Osler as a liberal champion—not a great stretch—and then, starting around 1900, let us survey the field he defended and assess the strength and intentions of his comrades and adversaries.

Osler flourished while medicine was emerging from an often unscientific and sometimes unsavory past. The nineteenth century had been stuffed with quacks and overrun with medical dogmatists following what Locke had long before called "that romance way of physick."[1] The relationship of the physician to his patient and to his colleagues and to science itself and the role of the physician in domestic and international society were controversies whose happy settlement was not uniformly anticipated. As Locke would have been, Osler was intensely interested in these issues.

Three historical processes, products of and complements to liberalism, came to bear on medicine in its early modern period.

The first was a gradual codification of the rules of professional medical conduct, generically referred to as "medical ethics," a term used rather carelessly[2] by Dr. Thomas Percival (1740–1804) as the title of a widely heeded book, published in Manchester, England, in 1803.

Percival's treatise was certainly about medicine but was not primarily about ethics; professional etiquette was the modal subject. Nonetheless, Percival described some ethically provocative duties, most remembered among them being the physician's obligation to violate statutory law if need be to assist a suffering patient. The problem under discussion was the treatment of men wounded while dueling, treatment then considered accessory to a felony. Percival said the doctor had to attend his patient, despite the law, "For in the offices of the healing art, no discrimination can be made, either of occasions or of characters."[3] An analogous obligation—to maintain most patient confidences even in defiance of state law or state order—was recognized in 1847 in the first Code of Ethics of the American Medical Association.[4]

Research ethics received new attention as well. Claude Bernard, in *An Introduction to the Study of Experimental Medicine* (1865), reviewed the subjects of animal vivisection, which he favored, and human vivisection, which, of course, he did not. Ironically, he listed Celsus among those who either practiced or approved human vivisection, confusing fair representation with active endorsement. Bernard asserted that "experiments," including innovative but unproven surgical procedures, were permissible within ethical "limits" which were not, in good practice, restrictive: "It is our duty and our right to perform an experiment on man whenever it can save his life, cure him or gain him some personal benefit. The principle of medical and surgical morality, therefore, consists in never performing on man an experiment which might be harmful to him to any extent, even though the result might be highly advantageous to science, i.e., to the health of others. But performing experiments and operations exclusively from the point of view of the patient's own advantage does not prevent their turning out profitably to science. It cannot indeed be otherwise."[5]

In the middle of the nineteenth century—during the slow rise of individual human rights but the rapid rise of utilitarianism (and long before the doctrine of informed consent)—he saw this issue exactly as Celsus had seen it: means *were* ends. Like many life scientists before and since, Bernard knew his heart better than his history.

The second historical process was the multinational maturation of a public-health ethic, exemplified by the ethics of the "medical democrats" in the German states and the "sanitarian" movement active in England and America from the 1830s onwards. The sanitarians, like William Farr and Edwin Chadwick in England, Lemuel Shattuck in Boston, and John Griscom in New York City, labored without confidence in the germ theory of disease but nonetheless recognized—and could prove statistically—that newcomers to growing industrial cities, especially to the poorer sections of those cities, were more likely to die prematurely than were the relatives and neighbors they had left at home on farms and in small towns.[6] Locke of the *Essay* would likely have located the cause of this increased mortality risk not in moral inferiority but in environmental decay and hygienic degradation. The sanitarians, though Victorian in tone as well as era, effectively did the same, arguing that health-compromising social inconveniences were problems of public circumstance which individuals could not hope to solve individually, and, since health was a right, governments had cause and duty to re alize solutions.[7] These ideas were radical and were not quickly or uniformly accepted. The American polity adapted to them slowly and imperfectly, well behind leading European societies. In the United States, the habits of immigrants and others of the lower and colored classes seemed a certain sign of *personal*-health and not *public*-health guilt,[8] and ambivalence about governmental social intervention of any seriousness betrayed first the fact and then the legacy of constitutionally sanctioned slavery and the relative federal impotence its protection had once required.

The third historical process was the rise of medical internationalism. In one respect, medical internationalism was ineluctable. Academic cosmopolitanism, imperialism, the European university system, and the breeding range of malaria-bearing mosquitoes made it so. But in another respect, medical internationalism was an invention.

In 1859, the Austrian army fought the French army at Solferino, and both forces left their wounded to die, for the most part. Jean-Henri Dunant (1828–1910), a Swiss citizen, tried to help these unlucky soldiers and others like them. By 1864, his efforts had led to the organization of voluntary relief societies in several countries and to the signing of the first Geneva Convention, by which instrument signatory governments obliged themselves to care for all military casualties, friend or foe. So began the International Red Cross. The United States joined the Convention in 1882, largely at the insistence of Clarissa Harlowe "Clara" Barton (1821–1912), schoolteacher, patent office clerk, and nurse-savior of American civil warriors. Having emulated for many Blue and some Grey what Florence Nightingale (1820–1910), a well-trained nurse, had lately done for the British Army in the Crimea, having helped to organize assistance for Franco-Prussian War victims while on a long "vacation" in Europe, and having founded the American National Red Cross as its first president in 1881, Barton entered the International Red Cross forcefully, soon expanding its mission to include natural-disaster relief.

The aims of the Geneva Convention and the aims of the International Red Cross advanced through many specific agreements, their common assertions adopted almost uniformly by the medical profession. Internationalism became more noticeably a mark of pride for medicine—and, we should say, for nursing—and became more frequently a point of honor for those physicians and nurses serving in, or conscientiously objecting to serving in, national armies. What had so long before made sense to Scribonius Largus now made conventions. And, in chauvinistic periods, controversies.

Osler tried to accelerate the evolution of his profession's ethical consensus, notably at the international level. In 1902, he addressed the Canadian Medical Association, meeting in Montreal. His subject was "Chauvinism in Medicine."

Osler opened his remarks—address and essay[9]—with a celebration of what he called "four great features of the guild": "*[i]ts noble ancestry*," its "*remarkable solidarity*," "*its progressive character*," and "its *singular beneficence*."[10] Regarding the third of these, he spoke proudly of "the unloading of old formulae and . . . the substitution of the scientific spirit of free inquiry for cast-iron dogmas."[11] Regarding the fourth, he saw "no limit to the possibilities of scientific medicine"; he cited but he did not endorse the hyperbolic projections of "philanthropists" and "philosophers": medicine as "the hope of humanity," medicine as the bringer of "Peace over all the Earth."[12]

Osler knew his profession's strength-for-good could be dissipated, its power misused, its promise of beneficence betrayed. Undermining the fourth great feature of "the guild" was a dangerous fault: "the name Chauvinism has become . . . [its] by-word, expressing a bigoted, intolerant spirit." There were two versions, according to Osler, "one . . . more apt to be found in the educated classes, . . . the other . . . pandemic in the fool multitude—'that numerous piece of monstrosity which, taken asunder, seem men and reasonable creatures of God, but confused together, make one great beast, and a monstrosity more prodigious than Hydra' *(Religio Medici)*." Either way, chauvinism was "a great enemy of progress and of peace and concord."[13]

Osler described the most damaging form of chauvinism, nationalism, as "the great curse of humanity." All modern life scientists should have acquired a robust immunity against nationalism "[t]hrough the International Congress and through the international meetings of the special societies," but they had not, and further correctives were indicated. More foreign study would help: "The man who has sat at the feet of [Rudolf] Virchow [German pathologist (1821–1902)] . . . can never look with unfriendly eyes at German medicine or German methods." And, beyond personal professional contact, "a knowledge of the literature of the profession of different countries will do much to counteract intolerance and Chauvinism." But internationalism in the medical literature, however praiseworthy, had not restored the Golden Age: "In the halcyon days of the Renaissance there was no nationalism in medicine, but a fine catholic spirit made great leaders like Vesalius [Flemish anatomist (1514–1564)], Eustachius [Italian anatomist (circa 1524–1574)], Stensen [Danish anatomist, physician, and geologist (1638–1686)] and others at home in every country in Europe. While this is impossible to-day, a great teacher of any country may have a world-wide audience in our journal literature, which has done so much to make medicine cosmopolitan."[14]

Why "impossible to-day"? Osler did not explain.

The neglect of Latin? If linguistic differentiation was the chief problem, then "a world-wide audience" would not have been available for "our journal literature."

Rigors of the road? Maybe. Osler eventually would "retire" to the Regius Professorship of Medicine at Oxford—the offer would come in a letter from British Prime Minister Arthur Balfour—largely to escape life in the railroad compartments of North America; over the twelve months of 1901, the year before "Chauvinism in Medicine," Osler had covered 27,000 miles in thirty-three professional trips, including two short ones to the White House.[15] An interstate practice, while trying, was not physically "impossible" in North America. Would an international practice have been physically "impossible" in Europe? The inconveniences of international consulting in Europe would surely have been less, especially less during the early twentieth century than during the Renaissance.

Perhaps Osler was speaking casually here. Perhaps he was just boosting "our journal literature," encouraging his audience to subscribe more cheerfully, to read more, to study harder. Nothing beyond that.

But Osler was discussing nationalism, a "curse" and a threat he did not underestimate. The old universal-science vision of Bacon and Boyle was as yet unrealized, and, as Osler surely knew and as we shall shortly observe, there were political-ethical differences among life scientists of different nations—particularly the Germanic and the non-Germanic nations—that were remarkable in their own day and would be incredible in ours.

Nevertheless, Osler's penultimate paragraph was sanguine. After all, it was 1902:

> With our History, Traditions, Achievements, and Hopes, there is little room for Chauvinism in medicine. The open mind, the free spirit of science, the ready acceptance of the best from any and every source, the attitude of rational receptiveness rather than of antagonism to new ideas, the liberal and friendly relationship between different nations and different sections of the same nation, the brotherly feeling which should characterize members of the oldest, most beneficent and universal guild that the race has evolved in its upward progress—these should neutralize the tendencies upon which I have so lightly touched.[16]

Osler was at once nationalistic, internationalistic, and transnationalistic, his patriotism the sentiment of a man who saw civilized order where others saw hegemony. Many of Osler's most distinguished German colleagues saw the world differently. Their political ethics, which we may fairly call perverse, had come even more conspicuously from the life sciences than had Locke's. But their political ethics, tragically influential, was "scientific" only metaphorically and analogously; it was in truth antiscientific, as we shall see.

The themes of the Montreal address reappeared in Osler's speaking and writing during the Great War, but in deeper relief. In October 1915, Osler traveled down from Oxford to address the Leeds University Medical School. His subject was "Science and War":

> The pride, pomp, and circumstance of war have so captivated the human mind that its horrors are deliberately minimized. . . . The inspiration of the nation is its battles. . . . For more than a century the world had been doing well—everywhere prosperity and progress. . . . An intellectual comity had sprung up among the nations. . . .
>
> And some of us had indulged the fond hope that in the power man had gained over nature had arisen possibilities for intellectual and social development such as to control collectively his morals and emotions, so that the nations would not learn war anymore. We were foolish enough to think that where Christianity had failed Science might succeed, forgetting that the hopelessness of the failure of the Gospel lay not in the message, but in its interpretation.

He cited Thucydides's dialogue between the Athenian delegates and the soon-to-be-slain citizens of Melos as "a parallel with Belgium" and as a prefigurement of disputes settled "in a twentieth-century 'might is right' fashion." He spoke on, sadly, bitterly, prophetically:

> In spite of unspeakable horrors, war has been one of the master forces in the evolution of a race of beings that has taken several millions of years to reach its present position. . . . Suddenly, within a few generations, man finds himself master of the forces of nature. . . .
>
> And what shall be our final judgement—for or against science? . . . To humanity in the gross, science seems a monster, but on the other side is a great credit balance—the enormous number spared the misery of sickness, the unspeakable tortures spared by anaesthesia, the more prompt care of the wounded, the better surgical technique, the lessened time in convalescence, the whole organization of nursing; the wounded soldier would throw his sword into the scale for science—and he is right.
>
> To one who is by temperament and education a Brunonian [a disciple of Sir Thomas Browne] and free from the 'common Antipathies' and 'National repugnances' one sad sequel of the war will be, for this generation at least, the death of international science. An impassible intellectual gulf yawns between the Allies and Germany, whose ways are not our ways and whose thoughts are not our thoughts. That she has made herself a reproach among the nations of the earth is a calamity deplored by all who have fought against Chauvinism in science.[17]

Sir William Osler, Baronet, "the Canadian Hippocrates," was an enthusiastic Empire man in the Great War, up until the death of his only surviving child in the Ypres salient in August 1917, a death despite the mercies of science, including a blood transfusion. But even at his most enthusiastic, as early as 1915, Osler had begun to mourn. Not yet for his second son, but for the unfulfilled prewar prospect of tolerant universalism. He saw that "mastery of the forces of nature" was already a problem, the "final judgment" of science already unsure, the "intellectual gulf" beyond which lay the German life-sciences community already "impassible."

Truly "impassible"? Not by war's end, not even for a liberal champion. In May 1919, in "The Old Humanities and the New Science," Osler addressed the Classical Association as its president. He set within an otherwise cheerful commentary on Greek and Latin authors a troubled argument with a terrible future: scientists could not fairly be held to wartime moral standards higher than those to which their fellow citizens might be held. This was a difficult case for Osler to make, and he made it awkwardly:

> [W]e have been accused of sinning against the light. Of course we have. Over us, too, the wave swept, but I protest against the selection of us for special blame. The other day, in an address on "The Comradeship of Letters" at Turin, President Wilson is reported to have said: "It is one of the great griefs of this war that the universities of the Central Empires used the thoughts of science to destroy mankind; it is the duty of the universities of these states to redeem science from this disgrace, and to show that the pulse of humanity beats in the classroom, and that there are sought out not the secrets of death but the secrets of life." A pious and worthy wish! but once in a war a nation mobilizes every energy, and to say that science has been prostituted in discovering means of butchery is to misunderstand the situation. Slaughter, wholesale and unrestricted, is what is sought, and to accomplish this the discoveries of the sainted Faraday and the gentle Dalton are utilized to the full, and to their several nations, scientific men render this service freely, if not gladly. That the mental attitude engendered by science is apt to lead to a gross materialism is a vulgar error! Scientific men, in mufti or in uniform, are not more brutal than their fellows, and the utilization of their discoveries in warfare should not be a greater reproach to them than is our joyous acceptance of their success.[18]

British physicians had early on tried to prevent retaliation-in-kind against Germany's use of poison gas on soldiers:

> What a change of heart after the appalling experience of the first gassing in 1915! . . . Surely we could not sink to such barbarity! . . . But martial expediency soon compelled the Allies to enlist the resources of chemistry. . . . A group of medical men representing the chief universities and

> medical bodies of the United Kingdom was innocent enough to suggest that such an unclean weapon . . . should for ever be abolished. . . . [W]e were given to understand that our interference in such matters was most untimely. All the same, it is gratifying to see that the suggestion has been adopted at the Peace Congress.[19]

Osler himself had tried to moderate Britain's response to the aerial bombing of civilians but found by the last year of the war that he had grown adequately accustomed to the practice to defend it as policy:

> With what a howl of righteous indignation the slaughter of our innocent women and children by the bombing of open towns was received. . . .
>
> Against reprisals there was at first a strong feeling. Early in 1916 I wrote to *The Times*: "The cry for reprisals illustrates the exquisitely hellish state of mind into which war plunges even sensible men. Not a pacifist, but a 'last ditcher,' yet I refuse to believe that as a nation, how bitter so ever the provocation, we shall stain our hands in the blood of the innocent. In this matter let us be free of bloodguiltiness, and let not the undying reproach of humanity rest on us as on the Germans." Two years changed me into an ordinary barbarian.[20]

Osler was not proud of his political-ethical metamorphosis and must have wished for its reversal. But the war-to-end-war was over, and, lest renewed hope now be squandered in recrimination, science quickly had to regain self-respect, self-confidence, and popular affection—on both sides of the old front.

Had he seen ahead to "science-and-the-*next*-war," would his faith in the cleansing power of genteel confession have been as sure? Not if he had feared sufficiently a threat still rising from within science itself.

Three Nonliberal Pathways

On February 17, 1858, in Berlin, in the second of twenty lectures on cellular pathology, Rudolf Virchow, building on the work of many before him, declared a new biological doctrine: "Where a cell arises, there a cell must have previously existed (*omnis cellula e cellula*), just as an animal can spring only from an animal, a plant only from a plant."[21] In August 1858, Alfred Russell Wallace (1823–1913) and Darwin jointly published evidence supporting a theory of species origination by a naturally selective process. Their effort was little noticed till the next year, when Darwin published his independent work as an "abstract," *On the Origin of Species by Means of Natural Selection, or the Preservation of Favoured Races in the Strug-*

gle for Life. In 1866, in a paper called "Experiments with Plant Hybrids," Gregor Mendel (1822–1884) published data supporting a theory of dominant and recessive genetic inheritance. Mendel's work effectively vanished till 1900 when three botanists, each one on his own nearing conclusions that were more or less Mendelian, "discovered"[22] the older research and conceded priority.

These events ensured disruption in the life sciences, in theology, in politics, and in ethics. A schism occurred in the political-ethical tradition of the life sciences, and three new, nonliberal pathways were cleared. Each led to a dead end, but each saw heavy traffic in its time, and each to this day sees an occasional misguided traveler. We shall look down these three pathways briefly in turn: cell-state theory, social Darwinism, and the eugenics movement.

Der Zellenstaat

The doctrine *omnis cellula e cellula* formalized a paradigmatic conversion in biomedical thinking. Questions of cellular function, tissue structure, organ-level coordination, and central regulation filled the intellectual agenda. The forms coming under finer scrutiny and the processes coming to better understanding were at once simpler and more intricate than previously realized. They were unmistakably functional, and, to scientists skeptical about the randomness of Darwin's evolutionary process, function easily implied purpose—as it had, both biologically and politically, to Aristotle. To describe that purpose, many leading researchers spoke of individual cells as organisms—not just single-cell organisms as organisms, which they are, but single cells *of* organisms *as* organisms, which they are not. Typical became Virchow's ill-founded conviction that "*[e]very animal presents itself as a sum of vital unities,* every one of which manifests all the characteristics of life."[23] Virchow and others spoke of tissues, organs, and whole organisms—"cell states"—in the language of political societies. "Cell" itself had among its meanings several that were faintly political; more so did other terms adopted by cytologists: "territory," "colony," "culture," "migration," "division of labor."[24]

Though nothing to Virchow beyond a convenience, heuristically helpful but philosophically insubstantial, the cell-state metaphor proved magnetically attractive. And not simply as a source of nomenclature. As early as 1859, Virchow was advising caution:

> What is an organism? A society of living cells, a tiny well-ordered state, with all the accessories—high officials and underlings, servants and masters, the great and the small. In medieval times it was customary to say that an organism was a microcosm, a little world. Nothing of the sort! The cos-

mos is no replica of the human being, nor is the human being a replica of the world! Nothing resembles life except life itself. The state can be termed an organism, since it consists of living citizens; conversely the organism can be termed a state, or a family, since it consists of living members of like origin. But here the comparison is at an end.[25]

Unhappily, not so. The cell-state metaphor persistently intrigued biologists trying to discover functional patterns in newly available data. A prominent pathologist proposed that disease represented a refusal of the cellular proletariat to participate in the communal work of the body. Class stratification was seen in the differentiation of central and peripheral nervous systems. The ganglion cell was likened to Friedrich Nietzsche's *Übermensch.*[26] In fairness to long-dead scientists, all innocent of the future, we should note that the veil of premodernity lifts cleanly from these metaphors on reading the modern sociobiological suggestion that "[t]he mind could be a republic of . . . schemata, programmed to compete among themselves for control of the decision centers" and that the "[w]ill might be the outcome of the competition."[27]

Clever, but indiscrete. And, in the old German case if not in the modern one, scientifically nonsensical, as shown by, among others, Hans Driesch (1867–1941).

Of independent means and temperament, Driesch worked outside the academy as an "amateur" embryologist. At first, Driesch's understanding of life was mechanistic; a cell was a "machine." Then, in 1891, he performed a most instructive experiment. Driesch split a sea urchin's fertilized ovum after its first division—at its two-cell stage. The two newly separated cells each then grew to become complete organisms, not "half-organisms" as expected. Observing this result, Driesch induced two hypotheses.

Driesch could not imagine a single "machine" that could divide itself into two new "machines" that were themselves complete and fully operational. Unprepared to consider this possibility on a molecular-genetic basis, Driesch speculated in Aristotelian terms: some external influence, some vital force—"entelechy"—must have been involved. By 1895, and to his ultimate discredit, Driesch had declared himself a "vitalist."[28]

Driesch's other hypothesis was much nearer the truth. At some point prior to its differentiation, a particular cell had an *indeterminate* "fate." Generalizing from the system under study (frog eggs behaved somewhat differently), Driesch inferred that the individual cells of multi-cell organisms were at some early stage pluripotent. Individual cells could somehow be induced to differentiate in one of *various* directions.

Could cells understood in this way ever again be considered "organisms"? No—kittens did not grow up to be dogs. Could pluripotent "citizens" function in a cell state? No—surely not in a Teutonic cell state; pluripotentiality implied social mobility.

Yet, even in a scientific age, and in no less than the leading community of that age, evidentiary challenge to *Zellenstaat* theory was ineffective. By the turn of the twentieth century, a cell-as-organism error was falsely but firmly established in German biology,[29] and the cell state was taken seriously as theory.

To many life scientists, especially to those abroad, *Zellenstaat* theory was an apostasy, a feudal conceit unworthy of its authors. Non-German critics were sometimes strident; German cell theory was "a rotten foundation" supporting "destructive" work, in the words of a Munich-trained English cytologist and protozoologist.[30] German critics as well could be forceful, but German critics attacked orthodoxy. German critics were dissenters.

In his 1907 textbook, *Plasma und Zelle* [*Plasma and Cell*], physician-anatomist Martin Heidenhain (1864–1949) could not avoid presenting *Zellenstaat* theory in detail; it was too widely accepted in German universities to be omitted. But Heidenhain, for one, had no trouble recognizing illegitimate epistemology, and its currency disturbed him:

> The doctrine of the cell state has become so popular that surely almost every lecturer in anatomy explains it in his presentations, and, indeed, according to my interpretation and sentiment, this doctrine has gradually lost the character of a mere comparison with the arrangements of human society, the character of a mere analogy or a metaphor, in that the true proportions of the biological constitution of the animal body are believed therewith to be expressed. The popularity of this notion is obviously based on the fact that it is difficult, where it is not impossible, to find a direct expression for the relationship of the near-order and sub-order of cells [for the relationship of "central" and "subsidiary" classes of cells] in the animal body, and so one has gladly and easily become habituated to the indirect paraphrase [*Zellenstaat* theory] on the strength of the comparison with a bourgeois state.[31]

Heidenhain went on to present *Zellenstaat* theory formally. He condemned the doctrine as "*[eine] Übertreibung des cellulären Prinzipes* [an exaggeration of the cellular principle]," discredited but pertinacious.[32] He then offered his own lengthy, systematic refutation.[33]

Heidenhain's voice was not entirely unaccompanied. When Driesch presented "a definitive statement of all that I have to say about the Organic,"[34] an attack on *Zellenstaat* theory was argumentatively prominent, if not climactic. But it was short—less than two pages in a transcript of nearly seven hundred. Driesch began the 1907 and 1908 Gifford Lectures at the University of Aberdeen by describing his intellectual "method." It was "to a great extent Kantian," he said, noting that Kant had built "on the foundations laid by Locke, Hume, and [Gottfried Wilhelm] Leibnitz [1646–1716]."[35] Driesch applied this "method" to the problem of teleology: were design, purpose, and intent properly found in biology? No, not in the sense that they were often sought by his colleagues. Driesch argued that destiny was an illusion in biology, and that it was likewise an illusion in human history, which itself was "nothing more than a branch of biological phylogeny in general . . . [,] not . . . a true evolution . . . but . . . a sum of cumulations."[36]

Driesch closed his 1907 lectures by attributing the discovery of biological truth to "the study of the living *individual* only." During the summer of 1908, he promised, his lectures would more fully explore what he called "the real philosophy of life, that is, the philosophy of the individual."[37]

Why such an exclusive emphasis on "the individual"? Termites, wolves, and *Homo sapiens* were social animals. History was "made" by social humanity and written, with few exceptions, by civilized humanity. Locke, the prevailing champion of the rights of the individual, had found in familial and societal dynamics explanations for many of the habits and failings of individuals. Locke's rejection of "innate ideas," a rejection cited by Driesch,[38] had left circumstance and teaching as the major shapers and misshapers of human behavior, and circumstance and teaching were largely social functions.

Driesch's individualist aim may all along have been predetermined—teleologically speaking—by his target, however distant that target must have appeared to his audience in Aberdeen. In 1908, as he had hinted he would do, Driesch attacked *Zellenstaat* theory directly. History, he said, revealed the operation of "no supra-personal factor":

> The *State* is *not* an "organism"—strange to say, for so very often in modern literature the real biological organism was pretended to be "explained" on the analogy of the State! Even the so-called "States" of bees and ants are real organisms only to a very small degree and not in detail.
>
> In order that any form of human society *might* properly be called an organism in itself, it would be required that disturbances of this organism should be repaired by force of the whole. But nothing of this sort exists: there *are* "regulations" in social life, as, for instance, when a business that needs workers attracts them by offering better payment, whilst an overly crowded business readily parts with work-people: but all this happens for the sake of the *individual's* liking and happiness, and for *no* other reason, as far as we know. There certainly is a little more of real organisation in the "State" of Hymenoptera [the order of membranous-winged insects to which bees and ants belong].[39] . . .

> As far as we *know*, the State—in the widest sense of this word—is the *sum* of the acting of all the individuals concerned in it, and is not a real "individual" itself.[40]

Though so clearly self-evident today that it may seem misplaced in the discourse of any scholar regardless of era, this argument had slowly become a marginal one in Imperial Germany. And its author, evidently, spoke from outside his own circle. Though he had gained his doctorate in 1889 under the preeminent physician-turned-zoologist, monistic philosopher, and Germanic activist Ernst Haeckel (1834–1919), Driesch was not "habilitated" as a professor until 1909, and, when the famous but still-unappointed Driesch presented his "definitive statement" over the summers of 1907 and 1908, he spoke in Scotland and wrote in English. Germany and German might not have proved as felicitous.

In the Second Reich, Driesch was politically out of step: "Thus this book [of lectures delivered in Scotland] may be witness to the truth which, I hope, will be universally recognised in the near future—that all culture, moral and intellectual and aesthetic, is not limited by the bounds of nationality."[41] In the Third Reich, Driesch would find himself unceremoniously a professor emeritus.

In the late nineteenth and early twentieth centuries, Germans led the life-sciences world, and a small number of "full" professors controlled the German life sciences. These professors were *Kulturträger*[42] [civilization bearers or culture bearers], and they were often men of political consequence. For example, Virchow, veteran 1848 revolutionary, "medical democrat,"[43] medical reformer,[44] and liberal member of Parliament, was co-founder of the Progressive party, and Oskar Hertwig and Wilhelm Waldeyer prominently supported expansion of the German Imperial Navy.[45] These men and their colleagues thought they saw through the lenses of their microscopes a working guide to large-scale organization—a timely observation during decades of national self-definition. The application of this guide to problems of political development seemed forward-thinking, both to themselves and to others.

Zellenstaat theory, like a chemical reaction, could be made "to run both ways." It could speak not only to the understanding of biological life but also to the understanding of political life. "Life" science could be a metaphor for "political" science, not just the other way around. Of course, Hobbes, long before, had derived a political ethics from "motion" and from physiological theories, and the Leviathan turned out to be a "cell state" in major respects; *Leviathan's* title page is well remembered for its giant sovereign in human-plated cuirass. But Hobbes's biological notions were retarded even for his own day, and his political advice never appealed much to his countrymen. Among the *Zellenstaat* theorists, by contrast, there were citizen-biologists of the first rank; neither their scientific teachings *nor* their political opinions were easily ignored.

Though composites were common, *Zellenstaat* theory had in retrospect two main political versions.

The "liberal" version was *descriptive* and communitarian—like Petty's seventeenth-century "political anatomy" but magnified and moralized. The human individual was a "commonwealth," and society was a "living" organism displaying collectively the efforts and the maladies of the individuals composing it.[46] Emphatically, there was a place for individual rights: "I uphold my own rights, and therefore I also recognize the rights of others," wrote Virchow in the preface to the first edition of *Cellular Pathology*. "This is the principle I act upon in life, in politics and in science. We owe it to ourselves to defend our rights, for it is the only guarantee for our individual development, and for our influence upon the community at large."[47] Virchow, indeed, would have been "quite at home among the English empiricists"[48] and attributed to Locke "all of the newer [progressive] movements in the State, in the Church, and in philosophy."[49] The "liberal" version of *Zellenstaat* theory was put forward by others as well, and in many ways, such as in Hertwig's 1899 address on the Kaiser's birthday: "*Die Lehre vom Organismus und ihre Beziehung zur Sozialwissenschaft*" ["The Doctrine of the Organism and Its Relationship to Social Science"][50] and in his 1922 book, *Der Staat als Organismus, Gedanken zur Entwicklung der Menschheit* [*The State as Organism: Thoughts on the Development of Mankind*].[51]

The "nonliberal" version of *Zellenstaat* theory was distinct; it was *prescriptive* and hierarchical,[52] xenophobic and aggressive. Society contained foci of inflammation and foci of foreign infection. These foci had to be found and managed, either by simple restorative procedures or by more radical means, including excision and killing *in situ*.

Both versions shared conceptual errors. *Zellenstaat* theory was unacceptably teleological and fundamentally prejudiced, nonempirical, nonscientific. The cellular doctrine upon which it was based was, like Newtonian physics, a terrific advance but not quite the whole story. *Omnis cellula e cellula* could not account for *first* cells nor for the differentiation of cells nor for the evolution of species. Differentiation and evolution (to the extent it was accepted by *Zellenstaat* theorists) were attributed to environmental influences. Thus was suggested the possibility of self-directed societal development—an appealing end, *assuming* an acceptable means.

In the history of political ethics, the key aspect of *Zellenstaat* theory was the interventionist function its "nonliberal" version assigned to the life scientist or physician. He had a socially therapeutic role beyond pursuit of knowledge, beyond ministration to individuals, beyond attention to public health in the modern sense. His contract was with society itself, not with rights-bearing patients one by one. The life scientist or physician was *eine Wanderzelle*, a migratory cell, immunocompetent and macrophagic, obliged to find and cure diseases in the body of the state.[53] As Driesch had innocently but prophetically remarked in 1908, any successful society-as-organism theory had to require "that disturbances of [the] organism should be repaired by force of the whole." Here was the force, soon to be used.

Zellenstaat theory persisted far into the twentieth century, long after its early intellectual respectability had faded. Of course, many, like Heidenhain and Driesch, rejected it. Forty-two percent of all German university teachers in medicine left the country after 1932;[54] some of these émigrés must have found repugnant their roles as state functionaries. Others were not so easily offended.

Social Darwinism

The disqualifying flaw in *Zellenstaat* theory should have been obvious all along: there was little reason to think that similarities between living cells and organisms on the one hand and states on the other were more than coincidental. A similar flaw should have been recognized in social Darwinism. Just as *Zellenstaat* theory compared society to an organism, social Darwinism compared societies and societal subgroups to *species* of organisms. Social Darwinism was less bizarre, less esoteric than *Zellenstaat* theory, but it was just as much a metaphor. It was also, arguably, a misnomer, as suggested in *The Descent of Man, and Selection in Relation to Sex* , wherein parts of Darwin's commentary on natural selection in civilized nations were comfortably Lockean:

> I have hitherto only considered the advancement of man from a semi-human condition to that of the modern savage. But some remarks on the action of Natural Selection on civilized nations may be worth adding. . . .
>
> We must remember that progress is no invariable rule. It is very difficult to say why one civilized nation rises, becomes more powerful, and spreads more widely, than another; or why the same nation progresses more quickly at one time than at another. . . .
>
> With highly civilized nations continued progress depends in a subordinate degree on Natural Selection; for such nations do not supplant and exterminate one another as do savage tribes. . . . The more efficient causes of progress seem to consist of a good education during youth whilst the brain is impressible, and of a high standard of excellence, inculcated by the ablest and best men, embodied in the laws, customs and traditions of the nation, and enforced by public opinion. It should, however, be borne in mind, that the enforcement of public opinion depends on our appreciation

> of the approbation and disapprobation of others; and this appreciation is founded on our sympathy, which it can hardly be doubted was originally developed through Natural Selection as one of the most important elements on [*sic*] the social instincts.[55]

Other parts of this same commentary were less comfortable and, as we shall see, less comforting, but, strictly on the matter of social Darwinism, Darwin himself was a nonconformist. Indeed, "social Spencerism," crediting the phrase, "survival of the fittest," to its true author, has been proposed as a fairer name.[56]

Herbert Spencer (1820–1903) was an English railway engineer and journalist-turned-philosopher who had guessed comparatively close to the mark on evolutionary theory as early as 1852.[57] In "The Social Organism," first published in 1860,[58] he considered the sociobiological analogy systematically and concluded that its heuristic advantages were attractive, its distortions tolerable. The body politic in Plato's *Republic* and the artificial man in Hobbes's *Leviathan* were famous early models of the social organism. *Leviathan* was the more compelling, its analogies the more detailed and "on the whole . . . the more plausible." However, the Leviathan, as a physioanatomical metaphor, was "full of inconsistencies." And no wonder. Prescientific perceptions of sociobiology were "necessarily vague and more or less fanciful. In the absence of physiological science, and especially of those comprehensive generalizations which it has but recently reached, it was impossible to discern the real parallelisms."[59] But "real parallelisms" did exist, and Spencer set out both to describe and to qualify them.

Spencer's social organism, at least in its original form, was not the focus of moral obligations beyond those idealized in an elementary social contract. In this respect, it was less audaciously conceived than Hobbes's Leviathan. Yes, the organism's "parts" needed each other; no, it was not a "Mortall God." But Spencer's analogy in another way went further than Hobbes's and passed boldly even the theory of the cell state, then in formation. Plato and Hobbes had compared circumscribed human societies to single human bodies. Instead, why not compare the *whole* of human society to the *whole* of biology?

Spencer imagined a taxonomy in which primitive tribes and "inferior" races took the parts of lower plants and animals. For example:

> The lowest animal and vegetal forms—*Protozoa* and *Protophyta*—are chiefly inhabitants of the water. They are minute bodies, most of which are made individually visible only by the microscope. All of them are extremely simple in structure; and some of them, as the *Rhizopods*, almost structureless. . . .
>
> Now do we not here discern analogies to the first stages of human societies? Among the lowest races, as the Bushmen, we find but incipient aggrega-

tion: sometimes single families; sometimes two or three families wandering about together. The number of associated units is small and variable; and their union inconstant. No division of labour exists except between the sexes; and the only kind of mutual aid is that of joint attack or defence.[60]

And so it was, all along parallel spectra of biological and social achievement till one found a genuine human being—or, at least, a vertebrate cerebrum—sitting in the Mother of Parliaments.[61] It is fair to assume that Spencer's Parliamentarian was some manner of Anglo-Saxon-Norman. Locke's, we should recall, could just as plausibly have been a Hottentot from Eton and King's.

According to Darwin, naturally selective pressure propelled the adaptive evolution of species, breeds, and races. Might it not simultaneously act within, among, and for or against societies? If so, reasoned social Darwinists in large and distinguished number, then much in the world that seemed either random or unfair was neither. Ruling classes, ruling races, and ruling nations ruled naturally—ruled, as it were, by "natural right." Egalitarian and redistributive reform—decolonialization, open immigration, unrestricted universal suffrage, antitrust laws, corporate and graduated income taxation—could only interfere with the vital process through which societies singly and collectively evolved to ever higher standards.

In Germany, ideas such as these fell into dangerous theoretical company. Ernst Haeckel saw correctly that in Darwinism no distinction had to be made between the natural selection of body and mind. An individual might prosper and procreate through strength or cleverness or swiftness or courage—or thickness of shell or timidity of manner or, simply, through good luck. But, in any event, structural and behavioral traits would find their futures *together*. So, "dualism," the separate consideration of flesh and spirit, was irrelevant, even antithetical, to the new understanding. "Monism" was dualism's necessary successor. And monism, as the word implied, was said to be greater than *and to contain* Darwinism. Indeed, Haeckel's intellectual goal was no less than the elaboration of an evolution-based nature-religion, the one and only true cosmic philosophy, and his particular (and, thus, ironic) interest was propagating among his countrymen its peculiarly German version. In 1899, Haeckel published his most influential nonscientific work, *Die Welträtsel*, a collection of easy monist essays—entitled in its 1902 American edition, one of many translations, *The Riddle of the Universe at the Close of the Nineteenth Century*. Body and mind were one; so were humans and their society; so, mystically, were a people, *ein Volk*, the transcendental mixed essence of nature and nation. Haeckel's titanic reputation as a biologist and the rhetorical force of his popular philosophical vision perennially commanded an audience, especially at home; by 1933, the year Adolf Hitler

(1889–1945) became chancellor, Germany alone was to have consumed nearly half a million copies of *Die Welträtsel*. In 1906, Haeckel established the German Monist League, a prosperous and openly antiecclesiastical group of forward thinkers and their many admirers, all dedicated to radical cultural change within the existing—the naturally selected—structure of social classes, each of which was properly to function like an organ within the body of the state. As romantic as it was scientific and passionately exclusive in its ethnic and racial politics, the League was ideologically an adumbration of National Socialism, the monist shadow appearing distinctly in Party teachings from historical determinism to eugenics. Haeckelian social Darwinism, in which evolutionary theory became a nationalistic pan-psychic quasi-religion, would have scandalized Darwin, and it was an amazement and an abomination to Haeckel's old teacher and former friend, Virchow.[62] In all but its modern European scientific packaging, and *exactly* in its cynical uses and terrible effects, it was a Teutonic Shinto. And, like its Japanese counterpart a generation later, it would recede with the empire it served. But it would also return renewed—after the Weimar years.

Many understood the risk posed by social Darwinism. Some, like the English physician and naturalist Thomas Henry Huxley (1825–1895), who once publicly assured an antievolution ecclesiastic that he would rather have descended from an ape than a bishop, also understood social Darwinism's logical error.[63] Many others did not. Outside the life sciences, critics often failed to differentiate between evolutionary theory and its misapplication, attacking good science and bad, either in confusion or by calculation.

Social Darwinism's most famous American opponent, the populist William Jennings Bryan (1860–1925), was not overly offended by evolutionary theory in general. In college, he had found Darwin's scheme engaging.[64] The descent of *Homo sapiens* specifically was something of a problem for Bryan, as it was, and is, for many Bible-believers: "I do not carry the doctrine of evolution as far as some do; I am not yet convinced that man is a lineal descendant of the lower animals."[65] Of course, the point was whether humans and lower animals had common ancestors. But, no matter.

Bryan, a religious conservative but a political liberal, felt the unpardonable sin of "Darwinism" was not its irreverence but its effect, its practical, *social* effect, its political-ethical effect, its tendency to excuse the exploitation of the weak by the strong, to excuse John D. Rockefeller, the British Empire, and the Kaiser.[66] "The Darwinian theory," Bryan charged, "represents man as reaching his perfection by the operation of the law of hate—the merciless law by which the strong crowd out and kill off the weak."[67] Again, as an aside, the wrong point.

As Woodrow Wilson's first secretary of state—he resigned to protest the handling of the *Lusitania* crisis[68]—Bryan became convinced that "Darwinism" underlay German militarism. By the early 1920s, he felt "Darwinism" had underlain the Great War as a whole and that "the menace of Darwinism" had to be checked on all fronts,[69] even at high personal and professional cost. Whence Bryan's revival-tent tours and the prefabricated spectacle of John Scopes's "Monkey Trial" in Dayton, Tennessee.

Many others, as well, understood social Darwinism's risks and effects. Those who simultaneously understood its logical error condemned it all the more fiercely.

In 1919, a two-volume *Festschrift* honored Sir William Osler on his seventieth birthday. The ninth of its 141 papers was "Social Reconstruction and the Medical Profession"[70] by Sir Auckland Geddes, a professor in a London medical school. Like Bryan, Geddes condemned social Darwinism as a factor in the late war. Unlike Bryan, however, Geddes distinguished sharply between evolutionary theory and its misapplication:

> We have all heard and wearied of the "Blond Beast." We have all groaned in spirit at the chorus of admiration raised by the Teutonic supermen to celebrate their own superiority. We have heard, some perhaps have believed, that their superiority was the inevitable outcome of their devotion to Science. I should write SCIENCE. To the world, the word Science denotes the sum of facts accurately known about the universe. To the Prussian, SCIENCE was almost synonymous with Truth, and SCIENCE was not only Prussia's rudder, but Prussia's chart. Now on that chart the Darwinian theory clearly could be marked only as a deep. In reality it is the most dangerous of all the reefs, for it appears to elevate to the dignity of a moral law a generalization based on observation of the processes of organic evolution and the elimination of the physically unfit, and from these draws deductions which are manifestly open to doubt. Applied to human conduct, it places each in antagonism to all, thus crowning self. This violates one of the most profound emotions of mankind, and denies the dearest instinct of the human race.[71]

Geddes conceded empiricism's corrosive effect on sacred tradition, but, unlike Bryan, he found the ascendancy of science over religion to be rich in moral compensation:

> During the short span of our generation physical and biological science have killed the emotions of religion and the Christian ethic in the minds of millions. Simultaneously medical science has helped to generate an emotion which is fast assuming the characteristics of religious thought . . . [in] the non-German world. . . . Perhaps we cannot define it yet, but we can see it at work as an emotion of human betterment in the great "uplift" move-

> ments, or can watch it finding expression in centres for child welfare, in schemes for housing the working classes, in the establishment of Ministries of Health, of Reconstruction and Research, in the growth of the Labour Party, in the spread of Socialism, of Syndicalism, and, incongruous though it may seem, in Bolshevism, and in the great ideal struggling to express itself through the League of Nations.[72]

The "emotion of human betterment" was not blunted by science. To the contrary. But the "emotion of human betterment," the compulsion for "uplift," could be turned to evil by a pseudo-science like social Darwinism.

Certainly, there *were* legitimate questions to be asked about the "evolution" of groups, races, cultures, and classes, and there *was* a role for the scientific method in answering them. How had Aristotle's "natural ruler" effected such an advantageous "union" with his "natural subject"?[73] How had Europeans come to dominate Native Americans, Africans, and Asians? Why had kidnapped sub-Saharan Africans survived longer on mosquito-infested alluvial plains than had the enslaved Native Americans they replaced? What made Prussians seem more warlike than Tahitians? Why did the working classes produce more violent criminals and fewer romantic poets than the landed gentry?

Nothing wrong with the questions. Just the methods and the answers, the former overly deductive, the latter typically dependent on presumptions of racial superiority, class superiority, and the like.

When clear-headed scholars and disinterested scientists—objective historians, well-disciplined anthropologists, empirical economists and sociologists, classical and molecular geneticists—finally explained away its theories, social Darwinism was revealed as self-confirming group delusion, as philosophical fraud, as political fakery. And those remembered as its leading spokesmen—Spencer and his fellow Englishman, the economist and journalist, Walter Bagehot (1826–1877); Haeckel in Germany; the American sociologist and political economist, William Graham Sumner (1840–1910); and numerous amateur adherents, including the cinema pioneer, David Lewelyn Wark "D.W." Griffith (1875–1948), director of *The Birth of a Nation* and apologist for slavery and its successor institutions in the southern United States; the celebrated aviator, Charles Augustus Lindbergh (1902–1974); and William Bradford Shockley (1910–1989), co-inventor of the transistor and Nobel laureate in physics—all declined in reputation, however well-meaning (or misunderstood, in Sumner's case) some of them might have been. Social Darwinism was theory in search of fact, theory in defiance of fact. The novels of Disraeli and Dickens and Conrad were trustier sources of sociopolitical insight.

The Eugenics Movement

Plato in the *Republic* had Socrates argue that the wise state would breed its citizens and would be justified in realizing eugenic goals through deception and through pregnancy termination or, failing that, infanticide.[74] Aristotle in the *Politics* defended a similar position, minus the deception and plus a distinction between lawful and unlawful abortion.[75]

For centuries, there had been no theoretical basis solid enough to support social engineering so audacious. Now, in the mid-nineteenth century, there seemed, finally, to be one.

Spencer concluded *The Principles of Biology* with a chapter called "Human Population in the Future." Evolution would continue, but in the improvement of which human characteristics would its naturally selective pressure be made manifest? Or, alternatively, toward the improvement of which human characteristics should an artificially selective pressure be applied? Within the first Darwinian decade, these discrete questions had already begun to fuse:

> Let us . . . consider in what particular ways this further evolution, this higher life, this greater co-ordination of actions, may be expected to show itself.
>
> Will it be in strength? Probably not to any considerable degree. . . .
>
> Will it be in swiftness or agility? Probably not. . . .
>
> Will it be in mechanical skill, that is, in the better-co-ordination of complex movements? Most likely in some degree. . . .
>
> Will it be in intelligence? Largely, no doubt. There is ample room for advance in this direction, and ample demand for it. Our lives are universally shortened by our ignorance. . . . [W]e have abundant scope for intellectual progress. . . .
>
> Will it be in morality, that is, in greater power of self-regulation? Largely also: perhaps most largely. . . . A further endowment of those feelings which civilization is developing in us—sentiments responding to the requirements of the social state—emotive faculties that find their gratifications in the duties devolving on us—must be acquired before the crimes, excesses, diseases, improvidences, dishonesties, and cruelties, that now so greatly diminish the duration of life, can cease. . . .
>
> No more in the case of Man than in the case of any other being, can we presume that evolution either has taken place, or will hereafter take place, spontaneously.[76]

What if natural selection failed to select "spontaneously" the best available traits, those "responding to the requirements of the social state"? What if the underclass scratched for survival more ruthlessly than the ruling class, or just bred more enthusiastically? Would not future society be benefited by the present-day limitation of breeding opportunities for those presumed genetically inferior? Social Darwinists might be right in the long run; leaving things alone might work well. But why risk it? And why wait? Signs of racial deterioration were already evident. Unlike other species, *Homo sapiens* could direct evolution itself, and had best do so, starting now.

Though he saw the hole in social Darwinism, Darwin himself, in excellent company, fell face-first in the eugenics ditch:

> With savages, the weak in body or mind are soon eliminated; and those that survive commonly exhibit a vigorous state of health. We civilized men, on the other hand, do our utmost to check the process of elimination; we build asylums for the imbecile, the mained [*sic*], and the sick; we institute poor-laws; and our medical men exert their utmost skill to save the life of everyone to the last moment. There is reason to believe that vaccination has preserved thousands, who from a weak constitution would formerly have succumbed to small-pox. Thus the weak members of civilized societies propagate their kind. No one who has attended to the breeding of domestic animals will doubt that this must be highly injurious to the race of man. It is surprising how soon a want of care, or care wrongly directed, leads to the degeneration of a domestic race; but excepting in the case of man himself, hardly any one is so ignorant as to allow his worst animals to breed.[77]

Darwin's views on social evolution, just pages away, seem contradicted by these muddy and admittedly borrowed[78] ideas. Was savage human vigor the result of short-term natural selection or individual training and experience? Was Darwin forgetting the robustness of feral animals, such as wild horses and wild dogs, and was he confusing signs of rough life, poor diet, and bad luck with evidence of heritable disadvantage? Was he forgetting the periodic need to import livestock to reduce the genomic risks of multigenerational inbreeding, whose usual purpose was to select traits irrelevant to or even *antagonistic* to genetic fitness, such as superfluous lactation in dairy cattle, docility—"bovine wit"—in draft animals, or equine traits in donkeys (*id est,* mules)? Darwin as eugenics conformer was not Darwin at his best.

Year after year, principals of the long-respected eugenics movement—Leonard Darwin, son of Charles, prominent among them—argued fervently and effectively for demographic activism. In *The Need for Eugenic Reform,* published in 1926, Leonard Darwin spoke for many well-meaning advocates of human racial progress:

> Our aim should be to substitute for the blind, cruel and often retrograde methods of natural selection some form of conscious selection which will make all men in future more able than at present to follow in the footsteps of the noblest human beings now on earth.[79] . . .

> Whatever be the kind of quality which it is held ought to be eliminated from the race, the following are the only methods which are available for that purpose: (1) The lethal chamber or murder, call it what you will; (2) Segregation . . . ; (3) Sterilization; and (4) Family limitation by conception control or abstinence.[80]

As a "method of racial purification," Leonard Darwin rejected the first option, but on consequentialist, not absolutist, grounds: too brutalizing for the murdering society.[81] Ultimately, as it happened, genetically "unfit" human beings in several technologically advanced countries would be gassed, segregated, or sterilized in the common interest and, it was often claimed, in *their own* interests.

This merging of interests, social and personal, was widely perceived and forthrightly described. In 1927 in *Buck* v. *Bell*, Oliver Wendell Holmes, Jr. (1841–1935), associate justice of the United States Supreme Court, wrote for the majority of his brother justices and, most likely, for the majority of his fellow citizens.

Carrie Buck was a young unmarried mother who was said, like her child and like her own mother, to be feeble-minded. She and her family were therefore promising instruments with which to establish the constitutionality of the 1924 sterilization statute in the state of Virginia. Tests were administered, examinations performed, and opinions rendered, and a consulting expert eugenicist, who never saw any of the principals, described mother, baby, and grandmother as "[belonging] to the shiftless, ignorant, and worthless class of anti-social whites of the South." Carrie Buck and her mother may have been shiftless and ignorant, and they and the baby *were* white Southerners, as were their custodians, but, by modern standards, no evidence credibly supported a diagnosis of imbecility, genetic or acquired. In fact, the baby, Vivian, seemed smart to her teachers through the second grade of elementary school, after which she died of causes unrelated to her intellect. Carrie's sister, Doris, sterilized without her consent or knowledge during an "appendectomy" in 1928, did not learn till 1979 why she and her husband, one Mr. Figgins, had never become parents.

Prospectively, the case for suspension of Carrie Buck's reproductive rights cohered well enough. The superintendent of the Virginia State Colony of Epileptics and Feeble Minded ordered the sterilization of Buck, whose guardian *ad litem* duly objected, suing the superintendent and, in time, his successor, Mr. Bell.[82] The Supreme Court of Appeals of the State of Virginia decided that Buck was "the probable potential parent of socially inadequate offspring . . . and that her welfare and that of society [would] be promoted by her sterilization." Holmes, considering an appeal, agreed:

> We have seen more than once that the public welfare may call upon the best citizens for their lives. It would be strange if it could not call upon those who already sap the strength of the State for these lesser sacrifices, often not felt to be such by those concerned, in order to prevent our being swamped with incompetence. It is better for all the world, if instead of waiting to execute degenerate offspring for crime, or to let them starve for their imbecility, society can prevent those who are manifestly unfit from continuing their kind. The principle that sustains compulsory vaccination is broad enough to cover cutting the Fallopian tubes. . . . Three generations of imbeciles are enough.[83]

Oliver Wendell Holmes, Sr. (1809–1894), physician, poet, novelist, professor of anatomy and physiology at Dartmouth and professor of anatomy at Harvard, discoverer—in 1842, before Semmelweiss—of the contagiousness of childbed fever, and father of Justice Holmes, might have complained across the breakfast table, had he been alive. As noted in the *Brooklyn Eagle* and then as passed along for better-bred consideration by an editorialist in *The Literary Digest,* it was "a curious fact that [the elder Holmes] pointed out that genius—or stupidity—may jump three or four generations, and reappear in offspring. Thus the child of a feeble-minded person might inherit genius from a great-great-grandfather. 'But,' *The Eagle* went on, 'this chance, in the opinion of Justice Holmes, apparently is so slender as to be negligible in the broad view of human society.' "[84]

The younger Holmes closed with a preemptive counter to the "unequal protection" criticism his opinion was sure to elicit, and, in doing so, he showed just how radical an image of societal cleansing even society's most judicious could now casually entertain:

> But, it is said, however it might be if this reasoning [*Buck* v. *Bell* as court-made law] were applied generally, it fails when it is confined to the small number who are in the institutions named and is not applied to the multitudes outside. It is the usual last resort of constitutional arguments to point out shortcomings of this sort. But the answer is that the law does all that is needed when it does all that it can, indicates a policy, applies it to all within the lines, and seeks to bring within the lines all similarly situated so far and so fast as its means allow. Of course so far as the operations [forced sterilizations] enable those who otherwise must be kept confined to be returned to the world, and thus open the asylum to others, the equality aimed at will be more nearly reached.[85]

Though the one dissenting justice did not record his objections, a contrary argument for the plaintiff was reported. Its vision of the future must have seemed preposterously dismal in America in the spring of

1927, as the major leagues began the most splendid season in the history of whites-only baseball. If paraphrased for Europe, though, the argument for Carrie Buck's personal liberty would have been insufficiently frightful:

> If this Act be a valid enactment, then the limits of the power of the State (which in the end is nothing more than the faction in control of the government) to rid itself of those citizens deemed undesirable according to its standards, by means of surgical sterilization, have not been set. We will have "established in the State the science of medicine and a corresponding system of judicature." A reign of doctors will be inaugurated and in the name of science new classes will be added, even races may be brought within the scope of such regulation, and the worst forms of tyranny practiced. In the place of the constitutional government of the fathers we shall have set up Plato's Republic.[86]

Eugenicists were not crazy. Many were social progressives, at least as many to the left of center as to the right, and some were life-sciences luminaries: for example, 1912 Nobel laureate Alexis Carrel (1873–1944), who once shared a *Time* magazine cover with Lindbergh, his biomedical-engineering associate and ill-advised philosophical protégé. All the more peculiar, then, considering intellectual resources, was the sense of urgency in the movement, peculiar because the human evolutionary process science was to assist was measured on a nearly geologic time scale, urgency or no urgency. In September 1929, while attending the International Congress of Eugenics in Rome, C. B. Davenport of the Cold Spring Harbor Laboratory, president of the International Federation of Eugenic Organizations, dispatched to Benito Mussolini a memorandum of German authorship. Mussolini was urged to enact a eugenics program without delay: "Maximum speed is necessary; the danger is enormous."[87] The "danger," one guesses, was procreation among the unemployed.

The eugenics movement was scientifically well founded as to *method* only in the broadest possible sense. The crude breeding of plants and animals had surely been antecedent to civilization itself, and more refined techniques had become vital to modern agriculture. Humans could be bred the same way, in theory. Yet in nearly every specific instance, the methods of the eugenics movement were scientifically baseless. Strictly, the movement could not be called "scientific" at all because it was neither experimental nor systematically observational or investigational. But large-scale social interventions suggested by the movement still *implied* a scientific method. And this implicit "method" was outrageous.

For instance, in 1912 Dr. Agnes Bluhm advocated conscious inattention to gravidas *in extremis.*[88] It should have been self-evident that if millions of years of natural selection had not already made labor and delivery easy and safe, then ignoring the agonies of individual women dying in childbed could hardly have been expected to improve the racial stock any time soon. But eugenics activists were strangely blind to many elementary observations.

As an objectively scientific exercise in human breeding, Dr. Bluhm's suggestion would have been extrapolated as follows: first, assess the "quality" of the human species; second, allow all suffering gravidas and their distressed fetuses to die peripartum deaths; third, persist in this policy for several hundred years; fourth, reassess the "quality" of the human species; fifth, compare the "quality" of the preintervention species with that of the postintervention species, regressing out all extrinsic influences that might have operated in the intervening centuries.

An experimental alternative would have been the use of selected subject cohorts and selected control cohorts followed over several centuries of group-specific breeding. This approach would have required much less initial intervention but much more elaborate long-term intra-experimental monitoring.

More to the point, however, the following of small, selected experimental populations over time would altogether have missed the goal of the movement: immediate and permanent large-scale social intervention—based, sadly, on misunderstood evolutionary theory and thoroughly unsubstantiated pseudo-scientific social theory.

As to *motivation,* both scientific and moral, the eugenics movement was even less well founded. The presumption that "losers" in contemporary social competition were genetically disadvantaged was grossly overdrawn. And the movement's consequentialist ethic—utilitarian at best, elitist at base—could never pass enlightened examination.

The Marxist Anglo-Scottish-Indian geneticist, J.B.S. Haldane (1892–1964), disgusted that there were so many eugenicists and eugenics sympathizers among his colleagues, whom he had a well-documented habit of offending under the best of circumstances, offered in *Heredity and Politics* in 1938 a searing refutation of the arguments that had prompted the drafting of sterilization laws in a number of countries, especially the United States. And he was no more amusable on the socially Darwinistic matter of racial prejudice, then not only rife but also respectable:

> Take any of the more blood-stained pages of the Anglo-Saxon Chronicle, in which aldermen's heads fall like so many ninepins. Read it aloud to a friend, substituting Kavirondo for Kent, Mbonga for Eadfrith, and so on. Then ask him whether he thinks it even remotely possible that in five hundred years the descendants of these bloodthirsty savages will be building

some of the world's most delicate architecture and laying the foundations of a remarkably just and stable form of government. If he says yes, you will probably find that he is either a firm supporter of Christian missions or a Marxist.[89]

The Nonliberal Convergence

In time and place, the convergence of these three nonliberal pathways was unmistakable and unforgivable. In perfectly modern times and in thoroughly polite societies, the public-health ethic—the ethic of the "medical democrats" and the "sanitarians"—had become dysmorphic.

Officials of several nations forced the involuntary sterilization of the genetically "unfit," and, in Germany and German-occupied territories, hundreds of thousands[90] of the "unfit" were killed for the health of society. More notoriously, in World War II, German clinicians and researchers conducted hideously abusive and routinely lethal medical experiments on prisoners—men, women, children. The effects of freezing and high altitude were studied; malaria was induced and various remedies tested; mustard gas injuries were inflicted and treated in several ways; simulated battlefield wounds were inoculated with organisms of interest in studies of sulfanilamide antibiosis; bone, muscle, and nerve injuries were inflicted, regenerating tissue biopsied, and bone transplantation techniques evaluated; chemically processed seawater was administered as a sole feeding to starved subjects; "epidemic jaundice" was induced in studies of the etiology and prophylaxis of hepatitis; thousands of sterilizations were performed in an effort to develop a simple population-control method, such as X-irradiation, suitable for mass application; cholera, diphtheria, paratyphoid, smallpox, typhoid, typhus, and yellow fever were induced and their prophylaxis contemplated or attempted; dietary poisons and poison bullets were tested; incendiary-bomb burns were simulated and salved with experimental preparations; 112 Jews, selected on the basis of skeletal features, were "defleshed" for museum use; and so forth.[91] All for the good of the *Wehrmacht*, the *Marine*, the *Luftwaffe*, the *Reich*, the *Volk*.

The criminal physicians of National Socialist Germany—some of them, at any rate—were prosecuted in the first of the Nuremberg trials: *The United States of America v. Karl Brandt, et alii*, "the Doctor's Trial," "the Physician's Case," "the Medical Case." Though the defendants in this first proceeding have been less well remembered than Hermann Göring, Rudolf Hess, Alfred Rosenberg, Albert Speer, and company, the crimes of the "Nazi doctors" helped illustrate for jurists and public the full variety of horrors for whose perpetration higher officials were subsequently hanged or imprisoned.

In his final statement to the Nuremberg Military Tribunal, Brandt, formerly Reich Commissioner for Health and Sanitation and Hitler's personal physician, his "Escort Physician," tried to explain what had happened to the life sciences in his country. Karl Brandt and Scribonius Largus exactly nineteen centuries before had quite likely held equivalent positions. But they did not hold equivalent opinions. Scribonius defended the maximization of *individual* welfare, Brandt the maximization of *corporate* welfare:

> There is a word which seems so simple—order; and how colossal are its implications. How immeasurable are the conflicts which hide behind the word obey. Both affected me, obey and order, and both imply responsibility. . . .
>
> I know things that disturb the conscience of a medical man, and I know the inner distress that afflicts one when ethics of every form are decided by an order or obedience.
>
> It is immaterial for the experiment whether it is done with or against the will of the person concerned. For the individual the event seems senseless, just as senseless as my actions as a doctor seem when isolated. The sense lies much deeper than that. Can I, as an individual, detach myself from the community? Can I remain outside and do without it? Could I, as a part of this community, evade it by saying I want to live in this community, but I don't want to make any sacrifices for it, either of body or soul? I want to keep a clear conscience. Let them see how they can get along. And yet we, that community and I, are somehow identical. . . .
>
> The meaning is the motive—devotion to the community. If on its account I am guilty, then on its account I will be answerable. . . .
>
> There was war. In war, efforts are all alike. . . . Did not [a certain pastor] . . . say that I was an idealist and not a criminal? . . .
>
> Here I am, subject of the most frightful charges, as if I had not only been a doctor, but also a man without heart or conscience. Do you think that it was a pleasure for me to receive the order to permit euthanasia? For 15 years I had toiled at the sickbed and every patient was to me like a brother. I worried about every sick child as if it had been my own. . . .
>
> I am fully conscious that when I said "Yes" to euthanasia I did so with the deepest conviction, just as it is my conviction today, that it was right. Death can mean deliverance. Death is life—just as much as birth. It was never meant to be murder.[92]

Twenty-eight "Nazi doctors" who survived the Nuremberg process were much later interviewed by research psychiatrist Robert Jay Lifton. By Lifton's analysis, a "healing-killing paradox"[93] had oozed up from a feculent ethics of national therapeusis and deterministic racial evolution. Germany's revitalization had to be accomplished prominently through biological renewal. The German people had to become an "organically indivisible national community." German physicians who shied from their duties in the New Order displayed "the symptom of an illness which threatens the healthy unity of the . . . national organism." National Socialism was "nothing but applied biology."[94] To cure the nation's sickness, "*anything*" was permissible.[95]

Der Zellenstaat, social Darwinism, and the eugenics movement were theoretically independent but effectively compatible. Each seemed "scientific" to its admirers, among whom were some numbers of scientists. Yet each was critically nonscientific. The first and the second were fine examples of well-reasoned theory applied far outside legitimate bounds. The second and the third undervalued the effects of nurturing and environment on social variation. Both these errors had long before been identified and elaborately discussed by Locke in *An Essay concerning Human Understanding.*

Der Zellenstaat, social Darwinism, and the eugenics movement were not centered on the individual, as life-sciences liberalism was centered. They were centered on other political objects: state, society, class, race. They had within them nothing of Hippocrates, nothing of Celsus or Scribonius Largus, nothing of Bacon properly read, nothing of Browne properly updated, and nothing at all of Locke, Percival, Dunant, Barton, or the prewar Osler. Taken together, then, *der Zellenstaat,* social Darwinism, and the eugenics movement formed an alternative to the liberal political ethics of the life sciences. But a *false* alternative. And not false for lack of liberalism. False for lack of science.

The Nuremberg Code and the "Lockean Person"

In the years between the world wars and then in World War II abruptly, the ethical structure of the life sciences collapsed in great sections. The damage done by eugenicists and medical murderers demanded urgent but durable reconstruction. First repairs were accomplished in 1947 in what came to be called the "Nuremberg Code," a ten-point declaration of ethical standards for human experimentation extracted from the Nuremberg Military Tribunal's decision in *Brandt.*[96] The following year, the World Medical Association's Declaration of Geneva revised the Hippocratic Oath so that new entrants into the medical profession would commit themselves to distinctly liberal ideals: "I will not permit considerations of religion, nationality, race, party politics or social standing to intervene between my duty and my patient; . . . even under threat, I will not use my medical knowledge contrary to the laws of humanity."[97] These reconstructions, as far as they went, proved sturdy, and, in time, they came to support elaborate additions, at least in the West.

We should note that life scientists in the communist bloc were not greatly affected by these changes. They remained servants of party and state—and, often, personality cult, as in the Soviet Union, where the pseudo-scientific agronomist Trofim Denisovich Lysenko (1898–1976) was destroying genetics.[98]

But that was the East. In the West, life scientists saw less reason to fear their states than to fear themselves, and, in the shadow of Nuremberg, they grew decidedly introspective. By the late 1960s, they and their many critics had elaborated a distinct body of commentaries: "bioethics."

A roughly philosophical demi-discipline, "bioethics" covered many clinical and research issues, old and new. But, oddly, it rarely if ever covered *political* issues. This omission was an ironic one, since the most commonly employed structural element in both clinical and research ethics post-Nuremberg had been a *political* ethic: liberalism, Lockean and Kantian.

Kant, of course, produced a distinguished political ethics, but among its major sources was neither Baconian science nor clinical epistemology. Locke's political ethics, as we have seen, was substantially a function of its author's empirical orientation and clinical experience; it would prove uniquely compatible with the life-sciences enterprise. No wonder, then, that the Nuremberg Code and the Declaration of Geneva implicitly confirmed as the living subject of clinical and research ethics a direct descendant of the "Lockean Person": freeborn, rational, rights-bearing as a human being, rights-bearing across borders, rights-bearing in defeat and even in death, co-equal with any physician, entitled to opinions, entitled to enter into therapeutic compacts and to withdraw from them at personal discretion, entitled to hear the truth about his or her own condition and to decide autonomously about treatment, about participation in clinical trials, and, generally, about the future.

No exceptions were intended. Even the "Nazi doctors" were themselves to be considered "Lockean Persons." Lifton's local institutional review board, the Yale University Committee on Research with Human Subjects, itself typical of a class of post-Nuremberg innovations, required that Lifton document that each "Nazi doctor" had freely granted fully informed consent to be interviewed and that Lifton, as principal investigator, had guaranteed to each "Nazi doctor" the full range of Nuremberg Code protections. All richly ironic. As Lifton observed, "The requirement itself stemmed from the Nuremberg Medical Trial, and was therefore a consequence of the misbehavior of the very doctors I was interviewing or their associates."[99] Ironic, yes. And, more revealingly, routine.

No exceptions were intended, but one qualification was—and is—needed. Even while being installed as the subject of clinical and research ethics after World War II, the "Lockean Person" fully asserted title to an asset far older than life-sciences liberalism and in critical situations more valuable than self-determination: physician beneficence. The contractual relationship of physician and patient, correctly understood, does not crowd out the parentalistic relationship that must exist between physician and patient much of the time. Even the "Lockean Person" gets sick. Or suicidal. Even the "Lockean Person" needs a physician who will respectfully act in a patient's *presumed* interest, occasionally in a patient's *imposed* interest. Parentalism undisciplined by rights-based liberalism is dangerous, as was demonstrated by a legion of conscientious eugenicists. But parentalism does have its place in the life sciences, as it has in life.

The "Lockean Person" might have evolved over "philosophic" time even without a brilliant progenitor—and might as well have done so, given the tepidness of genealogical curiosity exhibited even by admirers. Still, lineage aside, the "Lockean Person" would surely have been recognized and welcomed warmly by Locke's professional successors.

VI

FROM ALTRUISM TO ACTIVISM

IN 1905, an Alsatian university professor, having decided to devote himself "to scholarship and the arts" till the age of thirty and to the service of humanity thereafter,[1] put aside overlapping international reputations as pipe-organ virtuoso, musicologist, philosopher, and revisionist Christian theologian to become a medical student. In 1913, fully qualified as a physician and surgeon and specially trained in tropical medicine, he with his bride took ship to French Equatorial Africa, steamed up the Ogowe River to a mission station at Lambaréné, and made in the wilderness a hospital.

In 1899, Albert Schweitzer (1875–1965) had published *The Religious Philosophy of Kant*; in 1906, *Geschichte der Leben-Jesu Forschung*, famous in English as *The Quest of the Historical Jesus*; in 1913, the year he left Europe, a second edition of *The Quest* plus a new essay, *Psychiatric Study of Jesus*, written, surprisingly enough, to satisfy a requirement for his medical doctorate. These works, each revolutionary, had nearly cost Schweitzer his long-anticipated place in the mission field. Though he had promised the Paris Missionary Society "that [he] only wanted to be a doctor, and that on every other topic [he] would be *muet comme une carpe* (dumb as a carp)," one member of the Society's examination committee had resigned to protest the appointment.[2] The examiners had not been sure that they had their man, but they had been sure that they knew who the man before them was. Yet, the examinee, like many an answer-filled philosopher, was a far-from-finished thinker. His writings and his sermons may have suggested little room for large doubts, but Schweitzer's worldview, his *Weltanschauung*, had not been fully lit in Europe.

Why had this exquisitely civilized man now moved his future to this thoroughly primitive place?

Schweitzer had long assumed the Christian encouragement of humanitarian service to be answer enough. He had been a lucky and happy man; others had been miserable, and he had felt a duty to serve them. "Only one thing was certain, that it must be direct human service, however inconspicuous its sphere."[3] Well, medical service was maximally direct, and medical service in the jungle was sufficiently inconspicuous—and surpassingly dangerous by the stay-at-home standards of grown-up child prodigies. He was certainly "affirming the world"; he was not running away, physically or philosophically or theologically. He was not "denying the world" as a good Hindu or Buddhist or as a lazy and hypocritical Christian might have done.

As it happened, Schweitzer's explanation *of* himself *to* himself was forced to a conclusion by the once-again violent rivalries of his far-away European kindred. He had long held the contrarian's opinion that civilization, despite its show of progress (of which he and his training were exemplary) was in decline—and had been so for over a century. The great philosophers of the Enlightenment (*die Aufklärung*) had seen the world anew and had sensed obligation on every front; the advancement of human rights was their greatest legacy. Their worldviews had forthrightly been optimistic and ethical, but those of their chief successors had become pessimistic and selfish. Philosophers of the nineteenth century had shifted the goal of social progress from "the rights of man" to the romantic indulgence of supermen in superstates. Whence the decay of civilization, Schweitzer thought. And whence the Great War.

From August till November 1914, the doctor and his wife, Alsatians in a French colony, were interned. From the second day of this period and during what time he could spare thereafter for nine years, he wrote *The Philosophy of Civilization*, a project begun as early as 1900[4] but since then set aside, a project that refined his thinking as much as displayed it:

> As I worked along, the connection between civilization and our concept of the world became clear to me. I recognized that the catastrophe of civilization stemmed from a catastrophe in our thinking.
>
> The ideals of true civilization had lost their power because the idealistic attitude toward life in which they are rooted had gradually been lost. . . .
>
> In spite of the great importance we attach to the achievements of science and human prowess, it is obvious that only a humanity that is striving for ethical ends can benefit in full measure from material progress and can overcome the dangers that accompany it. . . .
>
> The only possible way out of chaos is for us to adopt a concept of the world based on the ideals of true civilization.
>
> But what is the nature of that concept of the world in which the will to general progress and the will to ethical progress join and are linked together?
>
> It consists in an ethical affirmation of the world and of life.
>
> What is affirmation of the world and of life?[5]

Here, Schweitzer for months was unable to respond. He could not answer the question generally, and he could not even answer the question for his own special case. Yes, by being a medical missionary in Africa, he was "affirming the world" and "affirming life," but why was he doing so? If his motive had been ethical, as it should have been, what ethical rule was he now obeying?

I felt like someone who has to replace a rotten boat that is no longer seaworthy with a new and better one, but does not know how to proceed.

For months on end I lived in a continual state of mental agitation. Without the least success I concentrated—even during my daily work at the hospital—on the real nature of the affirmation of life and of ethics and on the question of what they have in common. I was wandering about in a thicket where no path was to be found. I was pushing against an iron door that would not yield.

All that I had learned from philosophy about ethics left me dangling in midair. The notions of the Good that it had offered were all so lifeless, so unelemental, so narrow, and so lacking in content that it was impossible to relate them to an affirmative attitude [the attitude that had led Schweitzer to medical work in Africa].

Moreover, philosophy never, or only rarely, concerned itself with the problem of the connection between civilization and concepts of the worldview. The affirmation of life in modern times seemed so natural that no need was felt to explore its meaning.

To my surprise I recognized that the central province of philosophy into which my reflections on civilization and the worldview had led me was virtually unexplored territory. Now from this point, now from that, I tried to penetrate to its interior, but again and again I had to give up the attempt. I saw before me the concept that I wanted, but I could not catch hold of it. I could not formulate it.

While in this mental state I had to take a long journey on the river. I was staying with my wife on the coast at Cape Lopez for the sake of her health—it was in September 1915—when I was called out to visit Madame Pelot, the ailing wife of a missionary, at N'Gômô, about 160 miles upstream. . . .

Slowly we crept upstream, laboriously navigating—it was the dry season—between the sandbanks. Lost in thought I sat on the deck of the barge, struggling to find the elementary and universal concept of the ethical that I had not discovered in any philosophy. I covered sheet after sheet with disconnected sentences merely to concentrate on the problem. Two days passed. Late on the third day, at the very moment when, at sunset, we were making our way through a herd of hippopotamuses, there flashed upon my mind, unforeseen and unsought, the phrase, "reverence for life." The iron door had yielded. The path in the thicket had become visible.[6]

Schweitzer's new ethic, reverence for life, became by stages a famous modern philosophy, and Schweitzer himself became, by many accounts, as famous as any living human being: Francis of Assisi and David Livingstone and Captain Nemo, all as one man.

Reverence for life was both stimulus and response. It motivated *The Philosophy of Civilization* and was all along its object. By 1923, Schweitzer had produced the first two parts of what was to have been a fully articulated philosophical ethics strongly political in avowed purpose. "The Decay and Restoration of Civilization" and "Civilization and Ethics" were to have been followed by "The World-view of Reverence for Life," a detailing of arguments outlined in Parts I and II, and then by a treatment of "the Civilized State." The third part was never delivered formally, though reverence for life is no less clear for the omission. The fourth part must now be projected from the first two and from a career of action and advocacy recognized—far from its finish[7]—by award of the 1952 Nobel Prize for Peace.

In the preface to the first English edition of *The Philosophy of Civilization*, Schweitzer described himself as "a modern physician and surgeon" whose "moral conception of civilization [made him] almost a stranger amidst the intellectual life of [his] time."[8]

Schweitzer would have felt in friendlier company among the Illuminati and didacticians of the seventeenth, eighteenth, and the early nineteenth centuries: Bacon, despite his "somewhat worm-eaten personality,"[9] Hobbes, Locke, Adrien Helvétius (1715–1771), David Hume, Adam Smith (1723–1790), Locke's young student, the Third Earl of Shaftesbury (1671–1713), Kant, Jeremy Bentham (1748–1832).[10] They had been concerned with the ethical development of individual men and women and with the moral quality of interpersonal, intrasocietal, and international relations. The philosophers who had come after them were mostly of a different sort: Johann Gottlieb Fichte (1762–1814),[11] Georg Wilhelm Friedrich Hegel (1770–1831),[12] Artur Schopenhauer (1788–1860), Friedrich Wilhelm Nietzsche (1844–1900).[13] Renouncing their duty as social pedagogues, they had indulged their preference for speculation and "succeeded in deceiving themselves and others with . . . supposedly creative and inspiring illusion."[14] Then, following Darwin and Spencer,[15] "the natural sciences, which all this time had been growing stronger and stronger, rose up against them, and, with plebeian enthusiasm for the truth of reality, reduced to ruins the magnificent creations of their imagination." This revenge of the naturalists was justified, but it was awkwardly and destructively accomplished. "Since that time the ethical ideas on which civilization rests have been wandering about the world, poverty-stricken and homeless."[16] It was Schweitzer's intention now to offer as a permanent base for their soon-to-be renewed and redirected activism a safe, inviting, and respectable address: reverence for life.

The Philosophy of Civilization was Schweitzer's search for altruism. Some had argued that the path from egoism to altruism was a rational one, others that individuals acquired interest in the general welfare through social teaching, still others that egoism and altruism were both inherent (and that altruism could reliably be emphasized through exposure to well-chosen incentives). But these arguments and all their variants and combinations missed the mark. Altruism so explained was altruism no more.[17]

Like Plato much earlier, Kant had argued that the impulse to ethics was supranatural. Schweitzer added this judgment to Kant's credit but was annoyed by the timidity that had followed it. Ethical duties were not the product of sympathy (as Smith had argued elaborately in *The Theory of Moral Sentiments*) and they were not "hypothetical," not dependent on their consequences, not dependent on the future. Rather, they were knowable and known *a priori*. Ethical duties, including the duty not to employ humanity as a means to any end, were binding without regard to practical effect; in Kant's familiar language, they were "categorically imperative." Unfortunately, the "Categorical Imperative" was *itself* a consequentialist principle, serviceable as a patch for the rights-deficit in utilitarianism: "*Act only on a maxim by which you can will that it, at the same time, should become a general law.*"[18] For Schweitzer, Kant was a disappointment, a methodologist more than a moralist, since he shied away from giving his ethics what it most needed, a core belief: "On the whole he does nothing more than put the current utilitarian ethics under the protectorate of the Categorical Imperative. Behind a magnificent façade he constructs a block of tenements."[19]

Benthamite utilitarians constructed whole cities of tenements—real ones—convinced they could achieve the altruistic without understanding it. Biological utilitarians were fascinated by the puzzle of altruism's evolution, but they were philosophically unprepared to extend their arguments; or, like Spencer, they were all too ready to try.[20] The best attempt at relating the egoistic and the altruistic, Schweitzer thought, had only recently been made—but had been ignored outside a professional circle. In 1897, Wilhelm Stern, a physician with a Berlin practice, published his *Critical Foundation of Ethics as a Positive Science.*[21] Three years later, to an assembly of scientific colleagues, he delivered a short revision: *The General Principles of Ethics on a Natural-Science Basis.*[22] Schweitzer recognized in Stern's philosophy the first formal statement of what would be the spirit, if not the specifics, of reverence for life:

> The essential nature of the moral, he says, is the impulse to the maintenance of life by the repelling of all injurious attacks upon it, an impulse through which the individual being experiences a feeling of relationship to all other animate beings in face of nature's injurious attacks upon them. How has this mentality arisen in us? Through the fact that animate beings of the most varied kinds have been obliged through countless generations to fight side by side for existence against the forces of nature, and in their

> common distress have ceased to be hostile to one another, so that they might attempt a common resistance to the annihilation which threatened them, or perish in a common ruin. This experience, which began with their first and lowest stage of existence and has become through thousands of millions of generations more and more pronounced, has given its special character to the psychology of all living beings. All ethics are an affirmation of life, the character of which is determined by perception of the dangers to existence which living beings experience in common.
>
> How much deeper does Wilhelm Stern go than did Darwin! According to Darwin, experience of the never-ceasing, universal danger to existence produces in the end nothing but the herd-instinct, which holds together creatures of the same species. According to Stern, there is developed by the same experience a kind of solidarity with everything that lives. The barriers fall. Man experiences sympathy with animals, as they experience it, only less completely, with him. Ethics are not merely something peculiar to man, but, in a less developed form, are to be seen also in the animal world as such. Self-devotion is an experience of the deepened impulse to self-preservation. In the active as well as in the passive meaning of the word the whole animate creation is to be included within the basic principle of the moral.
>
> The fundamental commandment of ethics, then, is that we cause no suffering to any living creature, not even the lowest, unless it is to effect some necessary protection for ourselves, and that we be ready to undertake, whenever we can, positive action for the benefit of other creatures.[23]

Stern did go more deeply than Darwin, but he also went less cautiously. His long and learned arguments, as compressed by Schweitzer, seemed a bit sloppy. Lest only plankton be considered moral, eating-to-live had to be forgiven, and, then, by extension, many expressions of territoriality, the great source of meanness and violence among humans, had to be excused. The "forces of nature" against which "animate beings of the most varied kinds" had come to protect each other were, in actuality, exerted largely by hard-to-see beings called parasites, such as the agents of malaria and tuberculosis, against which protections were few. And, though tendencies and tropisms would one day be described across broad provinces of the living world, and even among the nonliving parasites called viruses, the "psychology of all living beings" was hard to imagine—exceedingly so in reference to plants and protozoa.

That said, Stern's sense that the strands of life were elaborately and tightly interwoven and that the vital fabric they composed had a moral right not to be ripped capriciously to shreds was a sense that many life scientists would come deeply to share in the new century. Schweitzer may have been among the first. And he was, he thought, "the first among

Western thinkers"—Stern included—who recognized the "crushing result" of this insight.[24]

Progression from the egoistic to the altruistic could be explained and encouraged through observations such as those made by Stern, but the righteousness of that progression—or any moral progression—could not be proved. Unprovability was in the nature of ethics, and Schweitzer was not disturbed by it.[25] He now proceeded to the control of altruism and then, finally, to its content.

Schweitzer argued that the *control* of altruism should be a function of individual moral judgment. It was "impossible to succeed in developing the ethic of ethical personality into a serviceable ethic of society" since "society exerts itself as much as possible to limit the authority of the ethic of personality" and "wants to have servants who will never oppose it."[26] Here, Schweitzer and Hans Driesch and Martin Heidenhain and, oddly, William Jennings Bryan would have agreed:

> Even a society whose ethical standard is relatively high, is dangerous to the ethics of its members. If those things which form precisely the defects of a social code of ethics develop strongly, and if society exercises, further, an excessively strong spiritual influence on individuals, then the ethic of ethical personality is ruined. This happens in present-day society, whose ethical conscience is becoming fatally stunted by a biologico-sociological ethic and this, moreover, finally corrupted by nationalism.
>
> The great mistake of ethical thought down to the present time is that it fails to admit the essential difference between the morality of ethical personality and that which is established from the standpoint of society, and always thinks that it ought, and is able, to cast them in one piece. The result is that the ethic of personality is sacrificed to the ethic of society. An end must be put to this. What matters is to recognize that the two are engaged in a conflict which cannot be made less intense. Either the moral standard of personality raises the moral standard of society, so far as is possible, to its own level, or is dragged down by it.[27]

Schweitzer must have seen the Darwinian herd instinct and the social-Darwinian absolution of racism and colonialism and war as easily confused, if not conjoined. Herd instinct might be a reality, but to concede to it an ethical priority over transsocietal and panhumanitarian ethics, let alone over life-reverential ethics, was to abrogate individual moral responsibility and to ensure societal amorality. He would have no part of this error, even in its domesticated Benthamite form. Herd instinct had to be transcended. A socially progressive but nonsocietally determined altruism had to be advanced.

Schweitzer argued that the *content* of altruism should be a function of reverence for life.

How did he know this? Not by revelation, despite the Pauline flavor of his housecall-up-the-river epiphany. Any true philosophy had to be derived from individual contemplation:

> With Descartes, philosophy starts from the dogma: "I think, therefore I exist." With this paltry, arbitrarily chosen beginning, it is landed irretrievably on the road to the abstract. It never finds the right approach to ethics, and remains entangled in a dead world- and life-view. True philosophy must start from the most immediate and comprehensive fact of consciousness, which says: "I am life which wills to live, in the midst of life which wills to live." This is not an ingenious dogmatic formula. Day by day, hour by hour, I live and move in it. At every moment of reflection it stands fresh before me. There bursts forth from it again and again as from roots that can never dry up, a living world- and life-view which can deal with all the facts of Being. A mysticism of ethical union with Being grows out of it.
>
> As in my own will-to-live there is a longing for wider life and for the mysterious exaltation of the will-to-live which we call pleasure, with dread of annihilation and of the mysterious depreciation of the will-to-live which we call pain; so is it also in the will-to-live all around me, whether it can express itself before me, or remains dumb.
>
> Ethics consist, therefore, in my experiencing the compulsion to show to all will-to-live the same reverence as I do to my own. There we have given us that basic principle of the moral which is a necessity of thought. It is good to maintain and to encourage life; it is bad to destroy life or to obstruct it.[28]

How reverential should one be? As reverential as possible. Kindness should be universal: no nonvoluntary sacrifice of the individual to the collective welfare, no avoidable killing of animals, no avoidable cruelty to animals in scientific experiments and physiological demonstrations,[29] no destruction of nonpestilent insects, no unreasoned destruction of plant life.

Schweitzer, by persistently "modern" Western standards, was eccentrically considerate toward nonhuman life forms, though he was not ascetic, not vegetarian (and, giving plant life its due, not anorectic), not friendly to pathogens, not humorless. He must have been affected by the antivivisectionism celebrated by a range of intellectuals and artists, notably Richard Wagner (1813–1883). He might have been stimulated by John Stuart Mill's (1806–1873) offhand acknowledgment that interspecies altruism was a moral problem, even for utilitarians.[30] And he must have known that the sympathies he shared with Stern were shared to some extent by many other life scientists.[31] Still, he knew that with his home culture, if not with Hindu culture, he was out of step. Anyone upholding his reverence-for-life standard had to expect some abuse, at least for a while:

> He is not afraid of being laughed at as sentimental. It is the fate of every truth to be a subject for laughter until it is generally recognized. Once it was considered folly to assume that men of colour were really men and ought to be treated as such, but the folly has become an accepted truth. To-day it is thought to be going too far to declare that constant regard for everything that lives, down to the lowest manifestation of life, is a demand made by rational ethics. The time is coming, however, when people will be astonished that mankind needed so long a time to learn to regard thoughtless injury to life as incompatible with ethics.[32]

In an era of unembarrassed imperialism, of unremitting mercantilistic extraction and exploitation, and when no industrialized society had come any closer to respect-for-nature than European romanticism and Shintoist pananimism, Schweitzer's "time" must have seemed far off indeed. Yet, within half a century and within just a few years of his own death, "environmentalism" had become a political force. And, within a generation after that, the high value of "biodiversity" had become not only a scientific assumption and an economic argument but a moral assertion as well. Schweitzer would have been pleased.

But not satisfied: "Ethics are responsibility without limit towards all that lives."[33] A standard so high would have to be a standard often violated, even if *never* ignored. Schweitzer knew this, but he was uncompromising, and he was not interested in forgiveness-through-cleverness. No Benthamite "moral arithmetic," no calculation of costs and benefits, no proof of optimization could turn bad into good:

> The ethics of reverence for life know nothing of a relative ethic. They make only the maintenance and promotion of life rank as good. All destruction of and injury to life, under whatever circumstances they take place, they condemn as evil. They do not keep in store adjustments between ethics and necessity all ready for use. Again and again and in ways that are always original they are trying to come to terms in man with reality. They do not abolish for him all ethical conflicts, but compel him to decide for himself in each case how far he can remain ethical and how far he must submit himself to the necessity for destruction of and injury to life, and therewith incur guilt. It is not by receiving instruction about agreement between ethical and necessary, that a man makes progress in ethics, but only by coming to hear more and more plainly the voice of the ethical, by becoming ruled more and more by the longing to preserve and promote life, and by becoming more and more obstinate in resistance to the necessity for destroying or injuring life.
>
> In ethical conflicts man can arrive only at subjective decisions. No one can decide for him at what point, on each occasion, lies the extreme limit of possibility for his persistence in the preservation and furtherance of life.

He alone has to judge this issue, by letting himself be guided by a feeling of the highest possible responsibility towards other life.

We must never let ourselves become blunted. We are living in truth, when we experience these conflicts more profoundly. The good conscience is an invention of the devil.[34]

Schweitzer offered no "decision rule," no mechanism for the lessening of moral tension, and he was particularly resistant to the imposition of such a rule by society. Correspondingly, his attitude toward private property, while strongly communitarian, was in no way socialistic. Wealth was "the property of society left in the sovereign control of the individual." The wealthy individual should, through his own "absolutely free decision" contribute to the common good. He might try to increase his wealth so as to employ his neighbors gainfully, or he might give to the needy directly.[35] At Lambaréné, Schweitzer "required" that the needy pay a pittance for their care, unless they were unable to do so, but his goal seems to have been parentalistic rather than economic: preservation of self-respect in Africans needing European cures.[36] There was no right to charity, as there was in Locke; in its stead was an obligation to serve any and all.

What if one's fortune or talent or training made him responsible for an undertaking to which the welfare of others was attached? He could easily and repeatedly be forced to choose between absolute but incompatible duties. What then?

The more extensive a man's activities, the oftener he finds himself in the position of having to sacrifice something of his humanity to his supra-personal responsibility. From this conflict customary consideration leads to the decision that the general responsibility does, as a matter of principle, annul the personal. . . .

No course remains open to current ethics but to sign this capitulation. They have no means of defending the fortress of personal morality, because it has not at its disposal any absolute notions of good and evil. Not so the ethics of reverence for life. . . .

I can never unite the ethical and the necessary to form a relative ethical; I must choose between ethical and necessary, and, if I choose the latter, must take it upon myself to incur guilt by an act of injury to life. . . .

The temptation to combine with the ethical into a relative ethical the expedient which is commanded me by the supra-personal responsibility is especially strong, because it can be shown, in defence of it, that the person who complies with the demand of this supra-personal responsibility, acts unegoistically. It is not to his individual existence or his individual welfare that he sacrifices another existence or welfare [but to that] of a majority. But ethical is more than unegoistic.[37]

Could the ethical man or woman avoid guilt by avoiding responsibility? In Indian and Chinese philosophies as read, admired, and rejected by Schweitzer, yes; avoidance might even have been preferred. In reverence for life, an uncompromisingly activistic ethic, no. Engagement with the world had to be accepted. Better yet that it be sought.

Could responsibility have been accepted but guilt avoided by its diffusion into the fiction of a group-person, such as a hospital staff, a corporation, a military force, a nation? Again, no: we must act "not in the spirit of the collective body, but in that of the man who wishes to be ethical. . . . Thus we serve society without abandoning ourselves to it. We do not allow it to be our guardian in the matter of ethics." For Schweitzer as for Locke, society served many a moral purpose, but society needed the ethical tutelage and the constitutional restraint of its citizens; contrary to the views of Fichte and Hegel, it had no right "[to arrogate] to itself the dignity of an ethical teacher." That it had so routinely captured this right had proved a tragedy: "The collapse of civilization has come about through ethics being left to society." Karl Brandt could have read these words—in German and in time:

> We do our duty to [society] by judging it critically, and trying to make it, so far as is possible, more ethical. Being in possession of an absolute standard of the ethical, we no longer allow ourselves to make acceptable as ethics principles of expediency or of the vulgarest opportunism. Nor do we remain any longer at the low level of allowing to be current, as in any way ethical, meaningless ideals of power, of passion or of nationalism, which are set up by miserable politicians and maintained in some degree of respect by bewildering propaganda. All the principles, dispositions, and ideals which make their appearance among us we measure, in their showy pedantry, with a rule on which the measures are given by the absolute ethics of reverence for life. We allow currency only to what is consistent with the claims of humanity. We bring into honour again regard for life and for the happiness of the individual. Sacred human rights we again hold high; not those which political rulers exalt at banquets and tread under foot in their actions, but the true rights. We call once more for justice, not that which imbecile authorities have elaborated in a legal scholasticism, nor that about which demagogues of all shades of colour shout themselves hoarse, but that which is filled to the full with the value of each single human existence. The foundation of law and right is humanity.[38]

Schweitzer was not just indicting the state-mediated criminality that had interrupted his medical practice and ruined Europe. He also wrote with a postwar conscience and, prophetically, a pre-war concern. While materialism alone could not nurture a healthy society, material deprivation would destroy it. Technological progress was a good, not an evil, but it always tended to distract humankind from the hard work of spiritual progress; in a future conflict, fought with weapons of unexampled killing power, this distraction could prove catastrophic.[39] No security could be found in the wisdom of common statesmen, attracted as they were to the false practicality of *Realpolitik* (or "political realism"):

> The objection is raised that, according to all experience, the state cannot exist by relying merely on truth, justice, and ethical considerations, but in the last resort has to take refuge in opportunism. We smile at this experience. It is refuted by the dreary results. We have, therefore, the right to declare the opposite course to be true wisdom, and to say that true power for the state as for the individual is to be found in spirituality and ethical conduct. The state lives by the confidence of those who belong to it; it lives by the confidence felt in it by other states. Opportunist policy may have temporary successes to record, but in the long run it assuredly ends in failure. . . .
>
> We are therefore freed from any duty of forming a conception of the civilized state which accords with the specifications of nationalism and national civilization, and we are at liberty to turn back to the profound *naïveté* of thinking it to be a state which allows itself to be guided by an ethical spirit of civilization. With confidence in the strength of the civilized attitude of mind which springs from reverence for life we undertake the task of making this civilized state an actuality. . . .
>
> Kant published, with the title *Towards Perpetual Peace*, a work containing rules which were to be observed with a view to lasting peace whenever treaties of peace were concluded. It was a mistake. Rules for treaties of peace, however well intentioned and however ably drawn up, can accomplish nothing. Only such thinking as establishes the sway of the mental attitude of reverence for life can bring to mankind perpetual peace.[40]

It was the character of the societies making treaties, far more than the treaties themselves, that would bring or keep or end a peace, and the character of a society, as manifested by its habits, institutions, and ambitions, was a function of aggregated individual moral choice. The foundation of law and right was *humanity*, not the other way around.

Schweitzer had not escaped exposure to statist philosophy by being born in a province of periodically exchangeable allegiances. Nor had he escaped all influence of *Zellenstaat* theory by being of professorial age when he entered medical school; the sign of the cell state—communitarian version only, though—was seen on several of his pages.[41] Still, he had, somehow, known the difference between the reigning political-ethical systems of his culture and his profession and political ethics as it should have been. His was a notoriously independent mind and a discouragingly impressive character, and, for all his personal gentleness, he was a terribly demanding ethicist—but he knew well that he was. "All this sounds too hard," he said.[42]

Schweitzer set ethical sufficiency beyond human capacity, but not beyond human understanding. His goals were simple, his obligations undenied, his successes genuine, and his failures universal. He knew that he served all best by serving each alone, and, more clearly than any other contributor to the political ethics of the life sciences, he knew why. Schweitzer pushed life-sciences liberalism beyond philosophy to an intellectually secure activism from which it need never retreat. Yet, he probably would not have called himself a "life-sciences liberal," even had he heard the term. He would more likely have said he was an optimist.

VII

LIFE-SCIENCES LIBERALISM IN ABSTRACT AND COMPETITION

AS EARLY AS 1923, the year Schweitzer published the first two parts of his never-perfected masterpiece, *The Philosophy of Civilization*, what we have been anticipating as "life-sciences liberalism" was fully formed. It still lacked the political authority provided in 1947 by the handing down of the Nuremberg Code, and it still lacked the intellectual authority provided in 1948 by the opening of the Lovelace Collection and in 1960 and 1963 by the addition of the Mellon Donations. Even taking the last of these dates, life-sciences liberalism could by now have been described and redescribed many times. Had it ever been abstracted, it might have appeared, roughly, as follows:

> *Method.* The unbiased observation of status and action and the unprejudiced and nonteleological interpretation of experience are keys to the understanding of behavior, including political behavior, and are prerequisite to the proposition of sound political-ethical theory. Most reliable is a "clinical" method, historically fastidious, observationally acute, free of unwarranted conclusion, and plainly argued.
>
> *Assumptions and assertions: rights.* Human beings have and retain rights to their lives, and to what tends to the preservation of their lives, to their liberty, health, limb, labor, and rightfully acquired real property, and to their personal prerogative and dignity. Though sometimes they may willfully in part be forfeited through criminality, human rights are the same for every human being, male or female, in every human circumstance. However convincingly his or her predicament may suggest otherwise, each individual is freeborn and rights-bearing. Each individual is—or should be respected as if he or she were—rational and self-interested. No one exists for another's use. There are natural rights to charity and sustenance, to intergenerational generosity, and to circumstances allowing self-betterment; or, at least, there is a personal obligation to give and to assist as if there were such rights. All human life should by rights be revered; all nonhuman life should by rights at least be respected, if not revered, and its use for human survival, sustenance, advance-

ment, edification, or pleasure should be well reasoned and seemly and, in the case of nonhuman animal life, compassionate.

Assumptions and assertions: behavior. Human behavior is a function of natural tendencies, teaching, and circumstance, and is substantially affected by "the law of opinion or reputation." Differences of custom, culture, religion, and philosophy arise naturally. Differences of politics also arise naturally. Unreasonableness and prejudice are psychological dysfunctions, and those who manifest them are politically disabled; rehabilitation is usually possible. Two particularly troublesome dysfunctions, sectarianism and nationalism, are both signs of narrowness of mind.

General conditions. Differences of custom, culture, religion, and philosophy should be tolerated. Differences of politics also should be tolerated if possible. One's enterprise, whatever it may be, can be judged only by the good it intends. Humankind must be served by serving individuals. Clinical knowledge must be sought only with individual welfare in mind and individual consent in hand and only to advance a present or future work of mercy.

Internal conditions. All agreements are made among individuals and all acts are committed by individuals, regardless of the structure into which those individuals are incorporated. The state is a convenient invention, and the corporation a productive one, but each malfunctions in predictable ways; the need for prophylactic regulation, periodic reform, and occasional "reinvention" should be anticipated. The work of life scientists and physicians and nurses may complement the interests of a state or corporation but must never be made subservient to such interests. Professional ethical responsibilities exist beyond statute and may require the defiance of statute. Professional ethical responsibilities exist beyond corporate obligation and may require the subordination of corporate interests.

External conditions. The intricate interdependence of societies is normal and healthy; restrictions on the transnational contact of individuals and groups should be minimal. Deterrent defenses, if necessary, should be proportional to real physical risk. The sovereignty of states, however majestically portrayed, should not greatly discourage well-measured transnational intervention in the interest of human rights.

Conditions governing ownership and distribution. The common inheritance of humankind is dispersed fairly only from abundance and only by honest labor. Surplus, however honestly acquired in peace or in war, must in significant part be redistributed to the needy and to the young. The life scientist or physician or nurse studies for those who cannot study, as others labor for him or for her. The ongoing work of the life sciences constantly expands the common inheritance of humankind and constantly generates surplus whose fair and compassionate redistribution is self-justifying.

Attitude. Tractable problems should actively be sought and, if found, should faithfully and competently be engaged.

We have seen life-sciences liberalism in formation and in refinement and in rivalry, but, so far, in rivalry only with its unstable illegitimate siblings, and, even in those contests, it needed substantively more than the strength of its own arguments to prevail. It needed, literally, the defeat of its rivals in war.

How has it fared against respectable contestants?

To answer this question, we must change techniques. Up to this point, political understanding has served ethical analysis. In what follows, a new *ethical* understanding will serve *political* analysis.

We will begin by sketching, with the fewest and thinnest of strokes, a pair of formidable philosophical challengers whose names should already be familiar.

The political ethics of the life sciences, exhibited in this book as life-sciences liberalism, naturally opposes ethics of "obligation" and ethics of "optimization." These two sets of ethics have traditional sources, each carefully tended and productively managed.

The first opponent is now often called "political realism," its moral stream formed by the confluence of the ethics of the state and the ethics of the corporation, both of them "corporate" ethics.

Like all "realisms" in philosophy, political realism claims to see fact as fact, nothing more: fact unpreceded by essence, unparalleled by ideal, unshadowed by value, unchanged by longing. But, less like its philosophical cousins, political realism sees as "indispensable fiction," if not literally as fact, an idealized form of one special object. That object may be the nation-state or the corporation or some other institution, and its idealized form is the unitary actor. In political realism, this idealization arises naturally from the object itself, ultimately defining it as valuable, endowing it with rights and responsibilities, and requiring those who serve it to serve it well. Thus, "amoral" political realism, in seeming self-contradiction, generates a system of values, a system of ethics.

A fair formal review of the ethical thought of political realism would be lengthy, but rewarding. We would first read through Thucydides's (*circa* 471 until *circa* 396 B.C.) unfinished chronicle of self-defeating self-interest in the Peloponnese,[1] Augustine's (A.D. 354–430) fusion of historical process with heavenly purpose,[2] Aquinas's absolution of the state's requisite sin,[3] Marsilius's (*circa* 1290–1343) courageous proto-

Hobbesian elevation of nationalism above papal transnationalism,[4] and Niccolò Machiavelli's (1469–1527) none-too-amusing guide to political advancement.[5] We would next consider Hobbes's theory of obligation and his ethics of competition and, in turn, the darkly liberating assumption of international anarchy (and, by inference, intercorporate anarchy) drawn dubiously—carelessly—from his sly and beautiful work, *Leviathan*,[6] by appreciative if sometimes less-than-fully witting realists. We would then move through Jean-Jacques Rousseau's (1712–1778) ominous animation of the "general will"[7] to Hegel's glorification of the state,[8] Carl von Clausewitz's (1780–1831) politico-military expediency,[9] and the *Machtpolitik* [power politics] of a young, but not an older, Max Weber (1864–1920).[10] After passing a socially righteous "season of violence" with the American theologian Reinhold Niebuhr (1892–1971),[11] we would reassess the indispensability of Edward Hallett Carr's (1892–1982) "indispensable fiction,"[12] the group-person so famously described by Hobbes, the group-person served by statesmen, diplomats, soldiers, dutiful war criminals, political terrorists, civic officeholders, corporate directors, corporate managers, and the odd hospital administrator. We would spend the Cold War with George F. Kennan (born 1904), honor-bound to be amoral,[13] and Hans Morgenthau (1904–1980), fixed on "THE NATIONAL INTEREST."[14] And, finally, we would weigh the hearts of those "lesser Common-wealths in the bowels of a greater,"[15] those targets of unblushing adoration and unrelieved disparagement: corporations, the neo-Hobbesian "artificial persons" defended by Milton Friedman (born 1912), mercurial American economist and free-market romantic, whose "Doctrine" it has been that "The Social Responsibility of Business Is to Increase its Profits."[16]

Contributors to the realist ethical tradition have differed in training, temperament, intention, technique, and influence. Their particular views have also differed. They have disagreed on the nature of *Homo sapiens*, on the nature of men and women in groups, and on the nature of intergroup behavior. They have variously placed the source of international and intercorporate ethical incompetence in heaven, in the individual, in the social class, in domestic society as a whole, and in the intercorporate "system"—the theater of great and small states, great and small corporations, and all self-absorbed institutions. They have disagreed on questions of guilt and absolution, some seeing sin where others have seen only tragic or not-so-tragic necessity. And, yet, most contributors to the realist ethical tradition have agreed on three sequential points. First, that corporate entities seek their selfish advantage when and where they recognize it. Second, that corporate selfishness, even cruelty, is often prudential. And, third, that acts of corporate prudence are ethically sound.

The first point is subject to objective assessment: do states, corporations, and institutions act selfishly? Or, in microeconomic terms, do they act rationally? There is much positive evidence on this question. But there is much negative evidence as well, and *non*realist theories of political action have grown up around some of it. The problem is reflectivity: an assumption of rationality in outcome presupposes an assumption of selfishness in motive. But the corporate pursuit of selfish interests may be hard to verify, given the complexity of options that may arise during policy formation. Failure to choose the superficially most selfish option might be evidence of caution or carelessness or compassion or even stealth. The cautious insinuation of ethics into policy formation cannot be distinguished from far-sightedness. The famous phrase "Food for Peace" could variously suggest an economic-warfare policy, a shrewd bargain, a high-minded gesture, a justification of mercy by appeal to self-interest, or a combination of the four. Realists can look only at finished decisions and impute selfishness in those who made them; they cannot even differentiate reliably between group selfishness and personal selfishness. This first point in the realist consensus is overtly the product of teleological argument.

Is this first point analytically useful, whether or not its implicit argument is teleological? It is useful to the extent that, as an assumption, it propels realist theory. Is this first point at the same time ethically safe? As long as its heuristic purpose is remembered, it is safe. But if its description of "normal behavior" is allowed to become a declaration of "normative standard," then the ethical safety of this first point in the realist consensus is degraded. Could we substitute a superior initial assumption, theoretically similar but ethically safer? A superior initial assumption would have to emphasize the complexity of corporate action; a superior initial assumption would have to disallow the diffusion of ethical responsibility. A safer initial assumption would not be a realist assumption.

The second point in the realist consensus—that corporate selfishness, even cruelty, is often prudential—is politically the crucial one. It is still a descriptive point and retains the possibility that selfishness may need to be "sacrificed" to some other goal, even to self*less*ness. But it is also a *pre*scriptive point for those comfortable with an obligation-based and corporatism-regarding theory of political good. It offers prudence as a meta-ethic, and it serves as a goal subsuming ethical goals, a short and sturdy bridge to the third point in the realist consensus.

The third point—that acts of corporate prudence are ethically sound—is realism's version of the end-justified means. It misses the role of "means" as "intermediate ends"—which they are—but is saved from licentiousness by the internal safeguard of prudence itself and by Machiavellian injunctions against becoming hated and against defining corporate goals extravagantly. But it invites two of the major perversions against which modern realists have so often warned. It invites the fusion of corporate interest with high moral principle, and it invites the identification of group interest with general interest: "What's good for us is good for all."

This third point is frankly *pre*scriptive. As part of a working ethical theory, it is highly suspect. An argument for "prudence" is very likely to be a tendentious argument, since "prudence" is not an ethical goal in itself, not timeless, but a function of one or more other goals and always a function of circumstance. Ethical truth therefore collapses to ethical relativism. And ethical relativism, when combined with realism's diffusion of accountability into the group-person, can be highly indulgent.

Realist ethics avoids moral judgment on ordinary criteria by emphasizing the extraordinary nature of its ethical actors and the extraordinary nature of the circumstances in which they exist. Attributes of these actors and circumstances are readily abstracted from realist assumptions: nations are immortal, and corporations and institutions may have a "life" beyond *or beside* the lives of the men and women who compose them; nation-states, corporations, and institutions possess personality; they make decisions that are or that become *or that must be respected as* unitary or nonfactional in nature; they exist among rivals in conditions of anarchy and, so, amorality; and the actions and the "rights" of the great are important, while those of the non-great are proportionally less important. These assumptions compose what we shall call the "realist illusion." Two assumptions identified with liberal socioeconomic theory are, paradoxically, complementary: the amoral pursuit of self-interest is rational; and the generalized amoral pursuit of self-interest inevitably serves the general welfare. These latter assumptions compose what we shall call the "liberal enhancement" of the realist illusion.[17]

Realist ethics is, distinctively, both absolutist and consequentialist; it is an ethics of *obligation*—first, last, and almost always, obligation *of* the individual *to* the group. Good realists ensure, through "rational action," that their state survives, that their corporation or institution prospers. Good realists aggrandize the object of their loyal concern, and, classically, they act toward it and for it *as if it were* a unitary, permanent, self-interested, self-defending, rights-bearing entity, a group-person: "France" or "Islam" or "Mitsubishi Heavy Industries" or, perhaps, "The Federal Bureau of Investigation"—a group-person unable to rely on any help but its own and, yet, a group-person to whose final success the welfare of many friends is tied. Thus, people sincerely identifying themselves with a societal consensus favoring altruism at home and abroad and urging personal and domestic political morality might at the same time firmly contend that their nation's power or their corporation's profit or their institution's prominence should in at least some instances be pursued amorally.

The second opponent for life-sciences liberalism is, ironically, utilitarianism, the prototypical "greatest-happiness-of-the-greatest-number" societal ethics, a great "liberal" majoritarian and communitarian advance in its early days and the core of the public-health ethic ever since. We will speak casually of a variant: "domestic" utilitarianism, "domestic" to stress that all individuals affected by it have ethical "standing" and to distinguish it sharply from the "foreign" ethics of inwardly utilitarian but outwardly realist societies, such as imperial democracies or publicly held corporations.

Though still famously associated with its founder, Jeremy Bentham,[18] and its leading exponents, James Mill (1773–1836)[19] and John Stuart Mill, utilitarianism had antecedents in classical philosophy and electoral republicanism, including the rights-conscious kind envisioned by Locke; it resonated harmoniously with the liberal economic theories of Smith and David Ricardo (1772–1823); and it has endured uncounted elaborations and tedious consensual refinements. All on a path from benevolent despotism to egalitarian governance to "Philosophical Radicalism" to communitarian utopianism to social engineering to civic orthodoxy. Parts of utilitarianism have so securely captured progressive thought that their acceptance has become automatic: *of course* majorities (or pluralities) should prevail; *of course* the social welfare function should be maximized; *of course* cost-benefit analysis is morally instructive. Yet, despite many attempts, most memorably J. S. Mill's near-apostasy, *On Liberty*,[20] and his sanctioning of the "rule of conduct" by "sentiment,"[21] the accepted parts of utilitarianism—including the Platonic "mathematized" utilitarianism known, loved, and feared as microeconomics—have never gracefully regarded and cannot reliably respect individual rights and welfare. Whence the occasional "utility" (to steal a term) of laws-above-laws: the Universal Declaration of the Rights of Man (which Bentham thought an incomprehensible travesty), the first ten and several subsequent amendments to the United States Constitution, the Geneva Conventions, the Nuremberg Code. And whence the life-sciences liberal's persistent unease with the public-health ethic.

Utilitarianism is consequentialist, rather than absolutist; it is an ethics of *optimization.* Good modern utilitarians—or the microeconomists they hire—ensure, however they must, that their society's aggregate health, wealth, or happiness be simultaneously as great and as widely distributed as possible. "Act" utilitarians may hope to avoid the infringement of what some others (or they themselves privately) would call the "rights" of individuals; and they may hope that the gap they open between minimal welfare and median welfare will, by chance, prove to be a narrow one;

but they are *not* required by their choice of political-ethical method either to protect individual rights or to preserve individual welfare. "Rule" utilitarians, fearing *long-term* utility diminishment brought about by the weakening of absolute principles violated for *short-term* utility maximization, let a quasi-Kantian conscience be their guide—sometimes. Good modern utilitarians justify means through ends, trying not to see that "means" are themselves "ends" (and the other way around). Perspective is their confounding and the future their undoing.

Though differentiable, realist ethics and "domestic" utilitarianism are, undeniably, cousins. They both stress the rights of collectivities, not individuals, realist ethics stressing the "external" rights of a group, utilitarianism stressing the "internal" rights of the majority (or the plurality) of a society. We have seen them both before, and we have also seen their affiliation; there could have been no *Zellenstaat*, no social Darwinism, no eugenics movement without "domestic" utilitarianism, and there could have been no "Nuremberg defense"—no diffusion of guilt into the fiction of a group-person—without the ethics of political realism.

We will next discuss these two opponents informally, but in some detail, through records of their modern behavior. We will see them prevailing, again and again and quite easily, sometimes with almost no notice, sometimes almost naturally, as if their hindrance would have been a shame. Our purpose, though, will remain a dull one: philosophical exposition, *not* journalistic exposé. Often, our attention will be drawn only to the inconspicuous, however sensational its circumstance. There will be little "new" in what follows, save viewpoint, and the expository virtue of its temporal setting—the latter part of the twentieth century—will be recognized as familiarity, not peculiarity.

VIII

PROTECTING THE STATE

THE TWO WORLD WARS were followed by partial reconceptions of national-security ethics. World War II particularly affected the national-security role of life scientists; as we have seen, the Nuremberg Code and other postwar instruments circumscribed that role within certain "liberal" principles. But the world wars had less formal effects as well.

From World War I, the battlefield use of poison gas both by the Central Powers and by the Allies has been most remembered. But the introduction of tactics even less discriminating has proved even more troublesome. The German Imperial Navy waged unrestricted submarine warfare, sinking merchantmen and passenger ships. German Zeppelins dropped bombs on civilian sectors of London. The Allied navies blockaded German ports, causing widespread malnutrition among noncombatants. And the Allies punitively withheld food relief after the Armistice in the first of many embittered, and embittering, peacetime acts.

From the introductory conflicts of the 1930s and from World War II itself, an even sadder heritage can easily be recalled. Consider a sequence: an Ethiopian village, Madrid, Guernica, Shanghai, Warsaw, Rotterdam, London, Berlin, Coventry, Belgrade, Leningrad (or, before and since, St. Petersburg), Tokyo, Cologne, Hamburg, Osaka, Yokohama, Dresden, Hiroshima, Nagasaki—all civilian centers attacked from the air. So listed, these names provoke a generalizable set of analytical reactions.

First, ethical injunctions are effectively degraded gradually.

Second, argument for reciprocity and argument for retribution—two species of argument by analogy—are efficient escalatory mechanisms.

Third, technological imperative can frame ethical apology, and ethical apology can sometimes be self-sustaining. For instance, two prerequisites to acceptably safe and accurate daytime strikes against legitimate military and industrial targets in Germany were long-range fighter-escort aircraft and reliable bombsights. Their long absence made "necessary" and therefore "justifiable" the nighttime bombing of large, flammable targets, such as cities. The "justifiability" of countervalue targeting easily outlived its "necessity."

Fourth, ethical distancing is enhanced by various factors: the interposition of political agents or technical instruments; physical separation of actors from the subjects upon which they act; the darkness—literal or figurative—in which acts are committed; elitism among actors, such as among officers of the *Luftwaffe* or Bomber Command or the 509th Composite Group, host of Enola Gay; and habituation to acts similar to those contemplated.

Fifth, ethical distance is proportional to actual outrage.

Sixth, the more transcendent the good pursued, the more agreeable the evil accomplished.

Each of these observations on the progress of aerial bombing could have its counterpart, real or potential, in a consideration of the involvement of life scientists in international competition. Against this involvement, the Nuremburg Code has proved imperfectly protective.

At the Nuremberg trials, states newly victorious in a war of nearly unrestrained savagery represented their values and interests in the persons of judges appointed to a military tribunal. Unitary rational action—the realist hypothesis and the realist ideal—had already begun to disintegrate within several of these states, factionalism and conflicting domestic ambitions returning to prominence. Still, the states-as-individuals sitting at Nuremberg were quite convincingly arrayed in realist vestments. And yet there prevailed at the trials three ideas *at odds* with realist ethics.

The first idea was that international standards of conduct did exist, that international anarchy and international war were moral aberrations, not constancies or inevitabilities.

The second idea was that Germany as a nation-state was not to be tried and not to be punished as at Versailles. The victors this time were not considering the condemnation of a fellow state but were dealing with the individual miscreants whom that state itself had failed to restrain. This was a trial of Germans as individuals and as individual members of illegal conspiracies. True, individual crimes had often reflected a grotesquely distorted societal morality; and, true, illegal conspiracies had usually been encouraged or required by the state itself. But, in the cases of most professionals and in the cases of those who had been able to affect policy, the United Nations sitting in judgment refused to allow the diffusion of guilt into the fiction of a group-person. The standard of conduct was not that of peacetime, surely, but defenses based on reasons of state were not generally allowed.

The third idea was that life scientists and physicians were governed by an internationally uniform professional ethical standard sometimes obliging them civilly to disobey the instructions of properly constituted authority within their own states. Of course, defendants other than life scientists and physicians were condemned on comparable grounds. Even some German military officers were so condemned; they should have refused "illegal" orders, from whatever source. But many military officers who failed to disobey were sympathetically viewed. Portraying the professional ethical responsibilities of military officers as *international* responsibilities was not so easy. And not so safe.

In Military Tribunal No. 1, Case No. 1, the trial of the "Nazi doctors," *realists* told life scientists they had a responsibility to be *liberals*, and *nationalists* told life scientists they had a responsibility to be *internationalists*—even *trans*nationalists. Life scientists were not to be *Wanderzellen.* They were to be autonomous moral agents held strictly to account for dishonoring their professed ethical principles. They were to be as Scribonius Largus had described them centuries before: patriots perhaps but never state functionaries. To the extent that they violated this injunction, they placed themselves outside *international* law, just as if they were pirates, subject to prosecution in any jurisdiction.

The number of people killed by the "Nazi doctors" was far less than the number killed by the "Nazi scientists" of Peenemünde: Wernher von Braun (1912–1977) and associates, the creators, builders, and launchers of the spectacular A-4 rocket, better known as the V-2, for *Vergeltungswaffen Zwei* [Vengeance Weapon Two]. A V-2 delivered a ton of high explosives from the edge of space supersonically—silently—onto large targets: Paris, London, Antwerp. Over 4,000 were used, the last within two weeks of the war's end. But von Braun and his fellows faced no tribunal. Indeed, they became heroes of American rocketry, and von Braun was rehabilitated to the point of hosting in his own name Sunday evening television programs "suitable for the whole family": "Tomorrowland" installments of the "Disneyland" series.

Was the difference between the treatment of the "Nazi doctors" and the "Nazi scientists" a function of the difference between the political ethics of the life sciences and a less restrictive political ethics of the physical sciences and engineering? Certainly, physical scientists and engineers have had a much different ethical history and have long figured prominently in both the violent and nonviolent pursuit of national and private corporate interests. Different expectations must have translated into different perceptions of guilt.

But another factor may actually have determined events. The work of the "Nazi doctors" was not only a violation of the political ethics of the life sciences. It was also extraneous to the political ethics of states. It could not advance the national interest of any of the conquering powers.

What if the "Nazi doctors" had done work of value to the perceived security interests of a victorious state? Which ethics would have dominated policy?

Ignoring a New Standard

On August 20, 1947, at Nuremberg, a verdict was reached in Military Tribunal No. 1, Case No. 1. Sixteen German physicians were found guilty of conspiracy, war crimes, and crimes against humanity. Some were also convicted of membership in a criminal organization. The war crimes charged against these sixteen physicians comprised human experimentation without consent and without compassion.[1] Their work had been appended to a genocidal enterprise independent of and at times in competition with the wider German war effort, a genocidal enterprise for which Hitler wanted to be remembered favorably, a genocidal enterprise in any event unhideable at war's end. Seven of the sixteen German experimenters were hanged at Landsberg Prison on June 2, 1948; the others served prison terms.[2]

The hypotheses entertained and the methods employed by Brandt *et alii* fell so far outside the norms of investigational practice that interpreting their results can be disorienting. No legitimate producer or consumer of human research data is experienced in the evaluation of such exotica; understandably, most have recoiled from it. In the 1980s, some of the toxic-gas[3] and hypothermia[4] data were to prove marginally helpful in civilian applications, but, even then, the conditions under which these data had originally been generated would severely restrict acceptance.[5]

Ethics aside, the experimental work of the "Nazi doctors" was not distinguished. It advanced no broad concepts, and many protocols were of poor design. In fairness, most protocols from the 1940s, whether from Auschwitz or the Mayo Clinic, would fail to satisfy the epistemological expectations of workers in later generations; yet, even by the standards of the day, the "Nazi doctors" reached few conclusions that were both credible *and* unapproachable using isolated systems or lower-animal models or true volunteers or consecutive, consenting civilian patients.

And, fatally for seven of them, the "Nazi doctors" had neglected to develop either unique technical abilities or workable advanced technologies of interest to forward-looking national-security leaderships.

On May 6, 1947, while the case against the "Nazi doctors" was still openly being argued, a "top secret" cable from Tokyo to Washington relayed a proposal from Ishii Shiro (1892–1958), a physician and microbiologist and former lieutenant general in the Imperial Japanese Army. While studying in Europe from 1928 to 1930, Ishii, so he said, had become intrigued by the outlaw status of biological warfare, and he determined eventually to benefit his employer, the Army Medical Corps, by realizing the potential of bioweaponry. In 1932, attached to the radically led Kwantung Army, which had recently initiated Japan's war against the

Chinese, Ishii had begun to organize in Manchuria a comprehensive military microbiology program, with addenda. The foremost activity was "defensive research": the development and mass production of vaccines and the study of assorted wartime problems, such as frostbite. The smaller, the better hidden, the less successful, and the more bitterly remembered activity was "offensive research": the study of human and agricultural diseases and human injuries whose induction might be tactically desirable and the development of procedures and munitions able to induce them dependably. Both types of research, but more remarkably the latter, took advantage of an audacious human experimentation program unique outside Nazi-occupied Europe and surpassing in fastidiousness any scientific effort judged at Nuremberg: multiple separate disguised facilities; at least three thousand subjects studied and "sacrificed"; masses of prospectively obtained human-subject data establishing minimal-lethal-dose levels for militarily attractive microbiological agents; detailed descriptions of untreated course in various induced diseases, such as gas gangrene; records of autopsies performed on human subjects freshly killed at specific stages in experimentally induced infectious diseases; experimental data on noninfectious diseases and injuries, such as through-and-through freezing of living, attached limbs; stacks of fixed tissue specimens; and unrivaled experience in the design, testing, field application, and hostile use of bioweaponry.[6]

Initial reports of Japanese activities were proving accurate as to scope, the cable said. If Ishii was now to be believed, high empirical standards had yielded dependable data. Their interpretation and supplementation could be arranged simply enough: "Ishii states that if guaranteed immunity from 'war crimes' in documentary form for himself, superiors and subordinates, he can describe program in detail. Ishii claims to have extensive knowledge including strategic and tactical use of BW [biological weapons or bioweaponry] on defense and offense, backed by some research on best agents to employ by geographical areas of Far East, and the use of BW in cold climates."[7]

A restricted United States government memorandum discussed Ishii's experience and his proposal in detail. His data and materials, most of which he had safely evacuated in the last week of the war to areas unlikely to be controlled by the Red Army, were of high quality. Though the Soviets had already interrogated one of Ishii's colleagues in Siberia and had interviewed Ishii himself and many of his subordinates jointly with the Americans, documents and materials were almost entirely, if not exclusively, in American control. A public trial would give the Soviets full access. And "[t]he value to the U.S. of Japanese BW data is of such importance to national security as to far outweigh the value accruing from 'war crimes' prosecution."[8]

Fully aware that American prisoners of war might have been among the subjects sacrificed (and believing that they had been),[9] fully aware that future exclusively Soviet judicial and journalistic disclosures would have to be dismissed as communist propaganda (as in fact they were, at least tacitly, in 1949 and 1950),[10] fully aware that "[e]xperiments on human beings similar to those conducted by the Ishii BW group have been condemned as war crimes by the International Military Tribunal,"[11] the United States agreed to Ishii's proposal. Except that his immunity was never put "in documentary form." The War Department favored a written guarantee, but the State Department, for once, suggested informality.[12]

Japanese technology was well received in the late 1940s. In the United States, about four thousand scientists and technicians had been involved in bioweaponry research and development during the war. In Britain, about five million anthrax-laden "cattle cakes" had been stockpiled by 1945.[13] But Western efforts had been crippled by constraints on research and testing. Dr. Edwin V. Hill, Chief, Basic Sciences, Camp Detrick, Maryland, writing in December 1947, stressed the Japanese advantage, noting the bargain price at which that advantage was being "Americanized":

> Evidence gathered . . . has greatly supplemented and amplified previous aspects of this field. It represents data which have been obtained by Japanese scientists at the expenditure of many millions of dollars and years of work. Information has accrued with respect to human susceptibility to those diseases as indicated by specific infectious doses of bacteria. Such information could not be obtained in our own laboratories because of scruples attached to human experimentation. These data were secured with a total outlay of Y [¥] 250,000 to date, a mere pittance by comparison with the actual cost of the studies.[14]

"It is hoped," he added, "that individuals who voluntarily contributed to this information will be spared embarrassment because of it."[15] Unrebutted 1986 congressional testimony charged that surviving American prisoners had been required to pledge—in writing and "under threat of court-martial"—to remain silent on the more sensitive aspects of their Manchurian captivity.[16]

Their crimes hidden behind a curtain of state, Ishii and his associates and thousands of project assistants quickly returned to private and professional life, often quite successfully. Some came to direct microbiological or medical institutions. Yoshimura Hisato, project director for the freezing and thawing of attached limbs in living humans, became a highly paid "freezing consultant" in the Japanese fishing industry and was first president of the Japan Meteorological Society.[17] Naito Ryoichi, another principal, established the soon-ubiquitous Green Cross Company, a medical supplier and pharmaceutical house, and became an oft-honored philanthropist.[18] Not until 1976 were the cruel achievements of "Unit 731" known in Japan; a sensational television documentary and a best-selling novel[19] finally brought some measure of disgrace where it was long overdue. Reports in the West came later.

With the human-experimentation problem "solved," American researchers were now able to make unexampled progress: "[Japanese] [d]ata . . . have proven to be of great value in confirming, supplementing and complementing several phases of U.S. research in BW, and may suggest new fields for future research. . . . This Japanese information is the only known source of data from scientifically controlled experiments showing the direct effect of BW agents on man."[20]

The American biological arsenal would soon feature several Japanese developments, including Ishii's "feather bomb," a device for spreading spore-forming bacteria, such as *Bacillus anthracis*, the agent of anthrax.[21]

What might have been the American policy alternatives in this case?

1. *Full public disclosure, a public trial by an international military tribunal, and genuine unilateral renunciation of all bioweaponry.* This would have been the only policy consistent with evolving Nuremberg principles and with the Geneva Convention of 1925, an instrument still unratified by the United States Senate in 1947.

2. *Full public disclosure, a public trial by an international military tribunal, and genuine unilateral renunciation of refinements in bioweaponry.* Under this policy, data of military interest would have become known to all participants, if not to the public at large. At worst, the Soviets would have chosen to copy advanced Japanese bioweaponry while realizing that other powers could have secretly been doing the same.

3. *Full public disclosure, a public trial by an international military tribunal, and unilateral refinement of bioweaponry, with or without public candor.* Condemnation of the Manchurian unit's crimes, followed by employment of its human experimentation data, emulation of its production techniques, and adoption of its tactical doctrine, would, obviously, have been hypocritical. But is hypocrisy a greater sin than imprudence? Not for the political realist. In 1947, there was no power, hostile or otherwise, suspected of possessing an offensive biological capability against which Ishii's weapons would have been needed as a deterrent. However, public disclosure and a public trial might have made it necessary to project an advanced near-term Soviet bioweaponry capability. So, an American decision to prosecute might logically have forced an American decision to refine its own arsenal, either secretly or openly.

4. *Limited public disclosure or no public disclosure, a token trial or no trial, and genuine unilateral renunciation of refinements in bioweaponry.* An overriding desire to deny Soviet access to Japanese advances might have led conscientious statesmen to the conclusion that the crimes of Ishii's group had to be ignored in large part or in whole, just as the better-known crimes of Italian politicians and warriors were for more ordinary reasons being ignored. However, since denying Soviet access to Japanese technology seemed feasible, a Soviet biological arms challenge based on Ishii's work could not have been thought likely. Genuine unilateral renunciation of refinements would have been the ethically logical move accompanying a decision not to prosecute.

5. *Limited public disclosure or no public disclosure, a token trial or no trial, and unilateral refinement of bioweaponry, with or without public candor.* The no-disclosure, no-trial, no-candor version of this option was the policy chosen, apparently by a near-consensus of informed officials. This choice was grossly inconsistent with the Nuremberg Code and could not have been defended as a response to an equivalent strategic or tactical threat. Unilateral advantage, the optimal incentive in realist ethics, was the unresisted attraction in this case.

What about the human experimentation data? Should they have been used, and, if so, should they have been used as they were used?

The idea that the "Lockean Person" was the only proper subject of human experimentation was just then being established at Nuremberg. The idea that individuals could ethically be made to suffer for the public good was still ambient. In the United States at the time, prisoners[22] and the feeble-minded[23] were widely thought to be usable in this way. So were the syphilitic African-American men tricked into joining and trapped into serving the Tuskegee Study,[24] a decades-long publicly funded observational research project whose protocol required that subjects be denied all antibiotics for whatever reason, for fear of obscuring the "natural history"—the untreated course—of their now too easily curable venereal disease. In the cases of prisoners and the Tuskegee subjects, and in many other cases[25] involving "ordinary" citizen-patients and respected investigators and revered institutions,[26] fictions of informed consent and voluntarism distorted ethical judgment. No such fictions could have covered Ishii's work.

On the other hand, Ishii's work in Manchuria was over. It was not ongoing, as the Tuskegee study was ongoing. All the subjects, or almost all, presumably, had been sacrificed at least a year and a half before. If the data drawn from their suffering were sound, why not use them? In the case of sacrificed American prisoners of war, might it not have been argued that all torments should finally be allowed to advance the national security, an outcome wholly consistent with the presumed wishes of faithful soldiers, sailors, and airmen? Learning from the case of a murder victim, after all, is not an endorsement of murder; in fact, such learning is legitimately part of a broad effort to save lives, especially the lives of those assaulted by would-be future murderers. Ishii's victims had been made to die in an epidemic of "biowarfare disease complex," a previously undescribed group of clinical entities. Why not learn from their records, from their necropsy reports, from their residual tissue specimens? Such an epidemic could recur.

There could hardly have been convincing absolutist arguments against the use of Ishii's results. And there could have been absolutist arguments for it; gas gangrene was a great killer of trauma victims in the days of restricted penicillin availability. But there should have been a contrary consequentialist argument that similarly unethical future practices might be harder to discourage if the data they generated could live in the body of citable science. The exemplary suppression of good data could have served a legitimate disciplinary purpose.

But these would have been the arguments of *research* ethics more than *political* ethics. The biggest problem with the use of these data was the *use* itself. These data were to be employed to facilitate the propagation of disease, to spread more widely and to spread intentionally the least discriminating effects and forms of warfare: sickness and starvation. These data were to be employed for an end far outside the bounds of the political ethics of the life sciences, though well within the bounds of realist ethics.

We should examine this last statement for signs of exaggeration. Ishii was himself a physician. Principals at Camp Detrick, later Fort Detrick, were life scientists of various types. Did not their behavior demonstrate the shallowness and pliability of the tradition we have exposed—or, worse, argue that we have exposed only one line of a terribly heterogeneous tradition? There are three answers to this question.

First, shallowness, pliability, and heterogeneity can be observed within many, if not within every, ethical tradition. As we saw in our discussion of *Zellenstaat* theory particularly, the liberal version of the political ethics of the life sciences was not nearly as prominent in Germany, for instance, as it was in the English-speaking world. Like their German counterparts, Japanese life scientists working in the 1930s and 1940s had not been nurtured in a liberal tradition.

Second, no ethical tradition can be held strictly to account for the actions of deviants. Ishii Shiro and his associates in the end sought to hide their past not only from the military justice of triumphant foreign powers but also from the censure of their own colleagues and countrymen.

Third, and most significant for our study, there was, at least on the American side of this deal, conscious subordination of a life-sciences ethical standard to a realist ethical standard. A choice between two roles was required and was made. "Life scientist as autonomous moral agent advancing the welfare of individuals" clearly lost, and "life scientist as state functionary" clearly won.

For all those who prevailed in World War II, surely, the crimes of the losers raised alarms. But of two kinds. After the trial of the "Nazi doctors," any invitation to elevate national interest above professional ethics could be seen as a comprehensive threat to both professional and personal integrity—and for life scientists most specifically. At the same time, and to opposite effect, reflection on the easy early progress of fascism and on the sinister habits of erstwhile ally Joseph Stalin could suggest that any *potential* threat to national interest should be seen as a *real* threat to national security, even to national existence. In any future industrialized conflict—in an even more sophisticated "wizard war," to use Winston Churchill's old term—the combatants would once again be societies themselves, scientists included.

While the crimes of the losers raised alarms, the crimes of the winners lowered inhibitions. If ever there had been means justified by their ends, such means could be found in the history of the late war. The fire-bombing of Dresden, though strategically superfluous, must surely have played some supporting role in the salvation of humankind, or so many believed. Reciprocity was prudent, retribution just. And capability had better be seen as intention, lest disaster befall.

The bioweaponry arsenal developed by the Japanese in Manchuria constituted a capability highly impressive to those privy to its existence. Though this capability had been acquired by the United States and denied to the Soviet Union, the Soviets soon showed themselves able to acquire, whether *de novo* or through espionage, a once-exclusively American capability far more impressive than an anthrax bomb stuffed with feathers. In August 1949, the Soviets exploded a nuclear fission device. Thereafter, the sophistication of any Soviet weapons program—such as a bioweaponry program—could hardly be doubted.

So much for Soviet capability. As for Soviet intention, events of the early Cold War were convincingly dreadful: civil war in Greece, coup d'état in Czechoslovakia, blockade on the access road to Berlin, revolution in China. Assuming the worst seemed wise.

The Army's Japanese-boosted bioweaponry program was a logical choice for acceleration. Bioweapons were non-nuclear, and, therefore, were candidate tools for a future "flexible response" strategy, and their further development would be technologically intensive and would thus play to a Western strength.

But employment of bioweapons was poorly conceptualized. Even vulnerability to bioweaponry attack—employment's less provocative mirror image—was poorly studied. The Army had only the previous year begun staging "biological field tests" simulating use of antipersonnel agents by a hostile power—or, of course, simulating use of the same agents by the United States.

The first series of these tests, conducted in Washington, D.C., used what was felt to be a nearly harmless bacterium, *Serratia marcescens,* an "aerobic" or oxygen-requiring, motile "bacillus" or "rod" that does not form spores and does not retain Gram's reagent—in usual parlance, an aerobic, motile, nonsporulating, Gram-negative rod.

S. marcescens is a "chromobacterium," so named for the red pigment elaborated by some of its strains. This red pigment deteriorates into a purplish-red viscous substance reminiscent of blood, a substance whose stunning appearance presumably explains many historical citations of supernatural intervention in the physical world. *S. marcescens* grows well on starch-laden foods, including, unfortunately, communion wafers.

Pigment elaboration and presumed harmlessness to humans long made this organism a favorite epidemiologic "marker." Spread on the hands of generations of microbiology students, *S. marcescens* conveniently demonstrated hand-to-hand, hand-to-nose, and person-to-person bacterial transmission. Sprayed into the air of hospitals and then cultured from surfaces, it tracked the routes of airborne in-hospital epidemics. Painted on the gums of patients awaiting dental procedures, *S. marcescens* could sometimes then be grown in cultures of blood drawn from peripheral veins, demonstrating postprocedure bacteremia. And so forth.

Like many other organisms once thought to be "saprophytic" and innocuous, *S. marcescens* is now recognized to be an important cause of serious infections in humans, particularly in debilitated, instrumented, or otherwise immunocompromised patients, typically in hospitals. *S. marcescens* infections in humans had in fact been reported prior to 1949, but rarely.[27] Most critically, at a point so early in the antibiotic era, the epidemiology of "nosocomial" or hospital-acquired infections and the pathophysiology of "opportunistic" infections were just beginning to be understood.

In fairness, then, *S. marcescens* must have seemed quite a good choice for the army's purposes, at least at first. From September 20 through 27, 1950, in the army's third set of secret field tests, *S. marcescens* was aerosolized from a ship off San Francisco. A second bacterium, also thought incorrectly to be harmless, *Bacillus globigii* or *subtilis,* a "simulant" for

Bacillus anthracis, the cause of anthrax, was also dispersed toward shore.[28] Air samples collected at forty-three different sites throughout the area were then analyzed for the presence of these organisms.[29] Useful data were collected, and all was assumed to have gone well.

Then, in October 1951, the *Archives of Internal Medicine* featured a paper by Wheat *et alii*, "Infection due to chromobacteria: Report of 11 cases."[30] This was an extraordinarily large series of patients, given the rarity with which *S. marcescens* infections were being recognized at the time. Indeed, the editor of the *Archives* paired Wheat's paper with a review article, "Bacteremia due to Gram-negative bacilli other than the Salmonella," in which *S. marcescens* was not even discussed.[31] Understandably, Wheat's paper was to become a landmark in the nosocomial literature.

All the infections reported by Wheat had been diagnosed at the same facility, the old Stanford University Hospital; all patients had had urinary tract instrumentation; all diagnoses had been clustered not only in place but also in time, eleven "rare" cases coming to attention within five months of each other, six within the first six weeks of the epidemic. These data strongly suggested introduction of *S. marcescens* into the hospital's "flora" shortly before the first diagnosis. To the army's dismay, Wheat's index case, ultimately a fatality, had been recognized September 29, 1950, nine days after spraying had begun and two days after it had ended.

General William Creasy, commander at Fort Detrick, asked four outside scientists whom he knew personally, each apparently linked in some way to the army's bioweaponry program, to comment on the Stanford problem. On August 5, 1952,[32] they reported back:

> 1. Experimental work in BW outside of the laboratory is impossible without the use of simulants. . . . An ideal simulant has not yet been found. . . .
>
> 2. Since the early days of bacteriology, SM [*Serratia marcescens*] has been the most commonly used organism for studying the dissemination of bacteria in air. Until recent years, there have been no reports of human illness associated with this organism in spite of its extensive use. . . .
>
> 3. The data in the referenced article [Wheat *et alii*, 1951] describing the experience in San Francisco are incomplete as to the primary relation of the SM isolated from the patients and their illnesses, except in the case of one patient who died from bacterial endocarditis and SM bacteremia. With this single exception, the finding of SM in these cases was not shown to have influenced the clinical course of the patients' illnesses.
>
> 4. On the basis of our study, we conclude that SM is so rarely a cause of illness and the illness resulting is predominantly so trivial, that its use as a simulant should be continued, even over populated areas, when such studies are necessary to the advancement of the BW program.

> 5. The program at Camp Detrick in the search for better simulants should be then actively pursued. If a more desirable simulant is discovered, it should then replace SM.
>
> 6. In future tests over populated areas, it would be desirable to institute prior and subsequent studies in a few hospitals to determine whether the report previously referred to was purely coincidental or whether the recovery of SM from patients was related to BW field tests. . . .
>
> In connection with open air testing, competent medical authority such as the USPHS [United States Public Health Service] stated no objection to the aerosolization of SM as a simulant test organism *under stated test conditions.*[33]

The four advisers suggested that Wheat's first co-author, a prominent epidemiologist, be "cleared" and asked to supply additional details, but the army never contacted anyone at Stanford.[34] In Senate testimony in 1977, an army staff officer reported that "the safety director at Fort Detrick did establish controls. His guidance was that SM would not be released over areas where debilitated or aged people were located, such as hospitals and sanitariums. He established rather stringent controls for pretests that were required in those areas before the bacteria Serratia marcescens could be released."[35] The first of these "controls" must have presumed that all "debilitated or aged people" were located in hospitals and sanitaria. It must also have presumed that hospitals and sanitaria were not located in populated areas of experimental interest. The second of these "controls" could not have been described as a protective measure, but only as a measure designed to increase the amount of data being generated while simultaneously improving their quality. In any event, the spraying of San Francisco remained a top army secret, as did further open-air research.

Biological field tests using live bacterial simulants of candidate antipersonnel agents continued until November 1969. Twenty series of these tests were staged in the "public domain" from 1952 until September 1968. *B. globigii* was now the favorite simulant, but *S. marcescens* was still used, appearing in three more series, the last in Hawaii in 1968.

We might note that open-air testing ended abruptly in the "public domain" 14 months before it ended on military bases. And that the final series was staged off the California coast at San Clemente in August and September 1968,[36] just as San Clemente's leading citizen was taking up his party's nomination for the presidency. On November 25, 1969, in what is remembered as one of his least self-interested acts, Richard Nixon (1913–1994) renounced America's bioweaponry program unilaterally.

The Army-*Serratia*-Stanford incident may well have been a coincidence. After much unwanted publicity beginning in 1976,[37] after United States Senate hearings in 1977,[38] and after a liability suit in 1981,[39] scientific consensus remained elusive. But our interest in this matter is not a function of serotypes and antibiotic sensitivities compared over a span of thirty years—an exercise in evolutionary biology as much as forensic pathology. Our interest is in ethical reasoning, and in this regard we need only acknowledge that the army genuinely but secretly doubted its own innocence in the Stanford epidemic.

Was the army—and were the life scientists working in or working for the army—violating the Nuremberg Code by engaging in human experimentation without benefit of informed consent? If so, was it the army's intention to be so engaged, or was the involvement of human subjects incidental, the intended subject being the wind? If human involvement was desired or was just unavoidable, were the human subjects *individuals* or was the human *subject*—singular—a population? Would such a distinction have made any difference, one way or the other? How would the ethics of this case have appeared if viewed through a utilitarian lens? Or without secrecy?

The ten-point research-ethics standard called the Nuremberg Code was included in the decision of the Nuremberg Military Tribunal in *The United States of America v. Karl Brandt, et alii,* handed down in 1947. All ten points applied to the army's testing program, but the first two points were particularly relevant.

The first point of the Code insisted absolutely that fully informed subjects had to grant truly voluntary consent prior to the initiation of any experiment in which they were involved. All those involved in the conduct of any experiment involving humans had "a personal duty and responsibility" to ascertain the quality of the consent granted.

The second point required that "[t]he experiment should be such as to yield fruitful results for the good of society, unprocurable by other methods or means of study, and not random and unnecessary in nature."[40]

Subsequent points enjoined against exposure to unnecessary or disproportional risk, identified incompetence with culpable negligence, and established minimum procedural safeguards.

An indictment of the army's conduct on grounds of defective consent would be uninterestingly easy. Before Wheat's paper was published, the army could have responded that the presence of humans in the populated areas being studied was incidental to its investigation—even that such presence was detrimental, that it was a hindrance—and that the innocuousness of the study design made its experimentation no more "human" than the launching of a weather balloon. After the Wheat paper, neither the "incidental" argument nor the "innocuousness" argument could have been used.

A more intriguing defense, one the army might have used even after the Stanford incident, could have been based largely on the second point of the Nuremberg Code. It would also have been consistent with Thomas Hobbes's statist ethic of irrevocable popular consent, with Reinhold Niebuhr's utilitarian ethic of internal group coercion, with Hans Morgenthau's realist ethic of the national interest, and with Paul Nitze's ambiguous ethics of the then-secret (but now famous) policy planning paper "NSC-68." The army could have argued:

- that its obligation to act "for the good of society," to use the Code's phrase, required it to assess the vulnerability of populated areas;
- that in assessing such vulnerability it was in no way experimenting on *individuals* but, rather, on a *population* with a presumptive interest in being defended;
- that the population of the studied area possessed no ethically unassailable mechanism for the mass expression of individual consent;
- that the benefits of the test were potentially great, given the dangers of the world situation;
- that the individual-based risks attending the test were small, if they were real at all, and they were greatly outweighed by the potential population-based benefits;
- that betraying the secrecy of the test prospectively or retrospectively would lessen its potential good to society by lessening its geostrategic value and would cause individual harm through panic and accident that would likely exceed all harm attributable to the test itself;
- that the data being sought were "unprocurable by other methods or means of study," to use another phrase from the Code.

In judging the attractiveness of such arguments in the early 1950s, we should recall the dual alarms raised for life scientists by World War II: the "threat-to-professional-integrity problem" and the "threat-to-national-existence problem." We should note with what facility realist arguments could fuse the concerns roused by both these alarms. Identifying moral integrity with national interest—the key identity in Morgenthau's most impassioned work—could compel a "nationalization" of professional ethical standards, especially in an era endangered by technologically driven risks. The realist appeal, as always, was not a seduction into amorality so much as a call to duty imposed by modern circumstance.

The army's bioweaponry program was a military enterprise. Its mission was nationalistic. It required life scientists but was not morally transformed by their involvement. If any moral transformation occurred, adoption of realist ethics by life scientists was the likely phenomenon.

If moral transformation in a similar setting could be made to run the other way, what arguments might achieve such a reversal? They might be arguments, or counterarguments, of two types.

Counterarguments based less selectively on the Nuremberg Code have already been outlined above. Counterarguments based more broadly on the political ethics of the life sciences would reflect a different view of the individual-corporate relationship and a different view of life scientists' responsibilities within and to the state. In their general form, these counterarguments would be:

- that humankind must be served by serving individuals;
- that knowledge must be gained whenever possible, but only with individual welfare in mind and individual consent in hand and only in the course of a work of mercy;
- that the work of life scientists may complement the interest of a state but must never be made subservient to such interest;
- that all agreements are made among individuals and all acts are committed by individuals, regardless of the structure into which those individuals have been incorporated; and
- that professional ethical obligations exist beyond corporate responsibility and may require the subordination of corporate interest.

Adapted to the specific situation above, these counterarguments would disallow nonconsensual individual human experimentation and nonconsensual population-based human experimentation, regardless of low risk or high purpose. They would also disallow the application of life scientists' efforts to projects serving the state in opposition to the individual, such as the development of a biological attack capability, but they would not disallow contributions to projects protecting individuals from the offensive designs of foreign states, such as the development of defenses against bioweaponry. Finally, they would require individual life scientists to understand the differences between illegitimate offensive research and legitimate defensive research, a difference so vague, by the army's own estimation,[41] as to make almost all conceivable biological warfare research illegitimate.

Similar in substance but distinctive in style was the search for pharmacological agents of national-security value, a search initiated by the United States Central Intelligence Agency (CIA) in the early 1950s. Project MKDELTA, established October 20, 1952, sought roles for such agents in clandestine operations. Project MKULTRA, authorized by Allen Dulles on April 3, 1953,[42] directed research into and development of specific agents potentially useful to MKDELTA, which project it contained along with related programs.

Much of what is publicly known about these undertakings has derived from MKULTRA's ten-year review, dated July 26, 1963, prepared by the Inspector General of the CIA and addressed to the Director of Central Intelligence, the DCI. The Inspector General's report and its covering memorandum, in large part declassified in United States Senate hearings held in 1975, identified political, organizational, and operational problems in both MKULTRA and MKDELTA and, by so doing, illuminated harshly, if indirectly, the conflict between the national-security imperative of realist ethics and the individual-rights imperative of life-sciences liberalism.

The Inspector General's report frankly exhibited the underside of realist ethics. It also exhibited three images fundamentally disturbing to life-sciences liberalism:

1. national-security managers and corporate operatives directly controlling life-sciences research and medical technologies;
2. national-security managers and corporate operatives establishing manipulative collaborative arrangements with reputable life scientists and leading universities; and
3. otherwise reputable life scientists willingly and at some level knowingly assisting projects *dis*reputable by the standards of their own profession.

Each of these images appears in the following extended extracts from the Inspector General's report. "Sanitizing" deletions from its original text were numbered consecutively: ". . . . . 6 . . . . . ," for instance. When they have fallen within passages of interest to our subject, these deletions have been duly "quoted."

MEMORANDUM FOR: Director of Central Intelligence
SUBJECT : Report of Inspection of MKULTRA

1. In connection with our survey of Technical Services Division, DD/P, it was deemed advisable to prepare the report of the MKULTRA program in one copy only, in view of its unusual sensitivity. . . .

5. MKULTRA was authorized by the then Director of Central Intelligence, Mr. Allen W. Dulles, in 1953. . . . Normal procedures for project approval, funding, and accounting were waived. However, special arrangements for audit of expenditures have evolved in subsequent years.

6. The scope of MKULTRA is comprehensive and ranges from the search for and procurement of botanical and chemical substances, through programs for their analysis in scientific laboratories, to progressive testing for effect on animals and human beings. The testing on individuals begins under laboratory conditions employing every safeguard and progresses gradually to more and more realistic operational simulations. The program

requires and obtains the services of a number of highly specialized authorities in many fields of the natural sciences.

7. The concepts involved in manipulating human behavior are found by many people both within and outside the Agency to be distasteful and unethical. There is considerable evidence that opposition intelligence services are active and highly proficient in this field. The experience of TSD [the Technical Services Division] to date indicates that both the research and the employment of the materials are expensive and often unpredictable in results. Nevertheless, there have been major accomplishments both in research and operational employment. . . .

I. Introduction . . .

2. The MKULTRA charter provides only a brief presentation of the rationale of the authorized activities. The sensitive aspects of the program as it has evolved over the ensuing ten years are the following:

a. Research in the manipulation of human behavior is considered by many authorities in medicine and related fields to be professionally unethical, therefore the reputations of professional participants in the MKULTRA program are on occasion in jeopardy.

b.

.....6.....

c. A final phase of the testing of MKULTRA products places the rights and interests of U.S. citizens in jeopardy.

d. Public disclosure of some aspects of MKULTRA activity could induce serious adverse reaction in U.S. public opinion, as well as stimulate offensive and defensive action in this field on the part of foreign intelligence services. . . .

c. Advanced testing of MKULTRA materials:

It is the firm doctrine in TSD that testing of materials under accepted scientific procedures fails to disclose the full pattern of reactions and attributions that may occur in operational situations. TSD initiated a program for covert testing of materials on unwitting U.S. citizens in 1955. The present report reviews the rationale and risks attending this activity and recommends termination of such testing in the United States, cf. paragraphs 10–18 below. . . .

II. Modus Operandi . . .

8. The next phase of the MKULTRA program involves physicians, toxicologists, and other specialists in mental, narcotics, and general hospitals and in prisons, who are provided the products and findings of the basic research and projects and proceed with intensive testing on human subjects. . . . The testing programs are conducted under accepted scientific

procedures including the use of control populations, the employment of placebos, and the detailed observation, measurement, recording, analysis, and publication of findings. Where health permits, test subjects are voluntary participants in the program. . . .

10. The final phase of testing of MKULTRA materials involves their application to unwitting subjects in normal life settings. It was noted earlier that the capabilities of MKULTRA substances to produce disabling or discrediting effects or to increase the effectiveness of interrogation of hostile subjects cannot be established solely through testing on volunteer populations. Reaction and attribution patterns are clearly affected when the testing is conducted in an atmosphere of confidence under skilled medical supervision.

11. TSD, therefore, entered into an informal arrangement with certain cleared and witting individuals in the Bureau of Narcotics in 1955 which provided for the release of MKULTRA materials for such testing as those individuals deemed desirable and feasible. . . .

12. The particular advantage of these arrangements with the Bureau of Narcotics officials has been that test subjects could be sought and cultivated within the setting of narcotics control. Some subjects have been informers or members of suspect criminal elements from whom the Bureau has obtained results of operational value through the tests. On the other hand, the effectiveness of the substances on individuals at all social levels, high and low, native American and foreign, is of great significance and testing has been performed on a variety of individuals within these categories.

13. A significant limitation on the effectiveness of such testing is the infeasibility of performing scientific observation of results. The Bureau agents are not qualified scientific observers. . . .

14. The MKULTRA program officer stated that the objectives of covert testing concern the field of toxicology rather than medicine; further, that the program is not intended to harm test individuals, and that the medical consultation and assistance is obtained when appropriate through separate MKULTRA arrangements. The risk of compromise of the program through correct diagnosis of an illness by an unwitting medical specialist is regularly considered and is stated to be a governing factor in the decision to conduct the given test. The Bureau officials also maintain close working relations with local police authorities which could be utilized to protect the activity in critical situations.

15. There have been several discussions in the public press in recent months on the use of certain MKULTRA-type drugs to influence human behavior. Broadly speaking, these have argued that research knowledge of possible adverse effects of such substances on human beings is inadequate, that some applications have done serious harm, and that professional researchers in medicine and psychiatry are split on the ethics of performing such research. Increasing public attention to this subject is to be expected.

16. The final step in the research and development sequence is the delivery of MKULTRA materials into the MKDELTA control system governing their employment in clandestine operations. . . . [I]t is appropriate here to note that the employment of MKDELTA materials remains an art rather than a scientific procedure. . . .

17. The final stage of covert testing of materials on unwitting subjects is clearly the most sensitive aspect of MKULTRA. No effective cover story appears to be available. . . . The handling of test subjects in the last analysis rests with the Narcotics agent working alone. . . . Suppression of knowledge of critical results from the top TSD and CIA management is an inherent risk in these operations.

18. Final phase testing of MKULTRA substances or devices on unwitting subjects is recognized to be an activity of genuine importance in the development of some but not all MKULTRA products. . . . Of more critical significance, however, is the risk of serious damage to the Agency in the event of compromise of the true nature of this activity. . . . A test subject may on some occasion in the future correctly attribute the cause of his reaction and secure independent professional medical assistance in identifying the exact nature of the substance employed, and by whom. . . . Weighing possible benefits of such testing against the risks of compromise and of resulting damage to CIA has led the Inspector General to recommend termination of this phase of the MKULTRA program. . . .

19.

..........17..........

. . .

III. Current estimate of the MKULTRA/MKDELTA capability

20.

..........18..........

. . .

27. Negative attitudes toward the use of MKDELTA materials; problems in the training of case officers in this field:

> The 1960 . . .27. . . report observed that some case officers have basic moral objections to the concept of MKDELTA and therefore refuse to use the materials. . . .

29. In summary, present evidence concerning the operational value of the MKDELTA capability would appear to confirm the principal judgments of the 1960 . . .31. . . report. There is an extremely low rate of operational use of the controlled materials. . . .

IV. Management of MKULTRA . . .

<u>It is recommended that:</u>

...............38...............[43]

Following the Inspector General's unfavorable review—a reiteration of previous criticism, apparently—and in light of his recommendations, whatever they were, the CIA reevaluated MKULTRA internally. In due course, the program was simply resubmitted for the Director's approval under a new name, MKSEARCH, its testing program retained with no recorded intention to limit subject selection to volunteers. The name-change scheme was advanced in a memorandum dated June 9, 1964, signed by Richard Helms, Deputy Director for Plans and himself a future Director of Central Intelligence.[44] Helms's memorandum included the following comment: "After eleven years of experience with the MKULTRA mechanism, it is my belief that the basic reasons for requesting waiver of standardized administrative controls over these sensitive activities are as valid today as they were in April 1953."[45]

One of the four under-signatures listed as "CONCURRENCES" was that of the same Inspector General whose recommendations Helms was now trying to modify.[46]

The Director evidently disapproved of further testing, but the balance of Helms's MKSEARCH proposal was funded through 1972. At least one drug experiment is known to have been performed under MKSEARCH; subjects were volunteers at the Vacaville State Prison in California.[47]

During the early postwar period—the early "Nuremberg Code period"—all parts of the United States defense-intelligence establishment, not just the CIA, were involved in secret drug testing and in drug administration for purposes incompatible with life-sciences liberalism. Drugs useful in interrogation settings were particular favorites. In one case, the army, Air Force, CIA, and FBI, all working through the Office of Naval Research, were able to disguise an interrogation-enhancement protocol as a motion-sickness study; it was performed on an unclassified basis at the University of Rochester.[48] Covert experiments using lysergic acid diethylamide (LSD) were also undertaken; some ended tragically.[49] And, as suggested over the years through the keyhole of journalistic and adversarial investigation and, in 1975, the limited commentary of a presidential commission and then as revealed spontaneously beginning in 1993 by a reformist secretary of energy, the general public, pregnant women, children, mentally retarded persons, and prisoners were all subject to experimentation, including exposure to radiation, that was misleadingly or falsely described or secretly conducted.[50]

In the stark case of MKULTRA, MKDELTA, and MKSEARCH, "realists" seemed so fully in character as to appear in caricature. The Inspector General's report was perfectly amoral. Ethical issues were recognized—manipulation of behavior, violation of individual rights and interests, abrogation of scientific standards in human research, human experimentation without consent, serious harm done to innocent and unsuspecting victims—but ethical issues were not ethical problems. They were political, organizational, or operational problems.

Protection of the country must have been the chief motivation for establishment of these projects in the early 1950s, and, though not referenced directly in the passages available, protection of the country must still have been a motivation ten years later. Parallel capabilities of foreign intelligence services were cited at one point in the declassified record, and some deference to the rights of United States citizens as compared to foreign citizens may have been displayed in the recommendation that unwitting subjects no longer be covertly tested—"poisoned" would have been the proper medical term—domestically.

But one wonders why Richard Helms was so intent that these projects continue minimally restrained when their products, developed at high human cost in some cases, were almost never used. It may be that realist ethics in the Inspector General's report and in Helms's memorandum was being most strongly expressed as *corporatist* ethics, the self-protective and self-perpetuating ethics of groups, conceptually the most extreme and surely the least romantic form of realist ethics. Protection of the Agency itself and the Agency's functions may inappropriately have become surrogates for protection of the country and may ironically have come to dominate those very values the Agency had been established to preserve.

The preceding three cases display arguments of realism overcoming arguments of liberalism easily, quickly, and repeatedly during the formation of national-security policies involving the life sciences. Other cases—or the same cases differently presented—could have displayed the opposite. Germany *did* allow Herbert Hoover to feed thousands in Belgium through neutral Dutch ports in World War I. The United States *did* allow Herbert Hoover, once again, to feed millions in Bolshevik Russia in 1921 and 1922. CIA agents did *not* often use the pharmacological tools made available to them. President Nixon, the mixture of his motives forgiven, *did* unilaterally renounce bioweaponry. But our task is not to keep score. It is to understand the highly dynamic tension between minimally compatible moral concepts and, thereby, to anticipate the risk posed by one to the other.

The incompatibility of life-sciences liberalism and political realism has been reconciled formally only once. The "Nuremberg Principle" rejected legal and ethical argumentation based on compliance with superior orders, formal or informal, overt or covert; the "Nuremberg Principle" established the individuality of moral responsibility in international legal tradition, if not in treaty or in domestic statute. The Nuremberg Code, an actual document, specified rules of clinical and research conduct; the Nuremberg Code implicitly confirmed the "Lockean Person" as full partner and prime beneficiary of medicine, guaranteeing human rights to patients and to research subjects and, thus, reestablishing liberalism in the life sciences. However convincing their resemblance to the unitary, rationally acting "leviathans" of realist theory, the states victorious in Europe in 1945 had concluded that clinicians and life-sciences researchers must always be liberals, that their moral agency must never be lost in the fiction of corporatism, that they must not subordinate professionalism to the obligations of "employment," military or nonmilitary, that they must be willing civilly to disobey their own states if necessary to protect the human rights of individuals.

All true. But if they had foreseen in the work of the "Nazi doctors" what they foresaw in the work of the "Nazi scientists" of Peenemünde, the states victorious in Europe in 1945 might have competed for custody. And the states personified as judges at Nuremberg might have mitigated their ruling, concluding that the crimes of the "Nazi doctors" had been crimes of a nation, crimes of an army, not crimes of individuals. Fortunately for international jurisprudence, Karl Brandt could not offer in trade the unimagined future of biotechnology.

The war-making potential of the life sciences was just barely worth considering at the close of World War II. The "Nazi doctors" had certainly not enhanced it. They were several intellectual strata below Germany's rocketeers and, unlike their American-shielded Japanese counterparts, they had not often shown themselves to be diligent investigators. They had little to contribute. Even the bioweaponry advances snatched from the path of the Red Army and offered successfully in trade in Tokyo were only incremental improvements, not fundamental transformations like the V-2, a medium-range ballistic missile just yearning for more thrust and a compact nuclear warhead. When bioweaponry's conceptual barriers finally began to crumble in the early 1970s, life-sciences research was still well outside the surveillance area of the realists writing international law.

On April 10, 1972, in Washington, London, and Moscow, there was concluded in ignorance of its true significance a new treaty: the "Convention on the Prohibition of the Development, Production and Stockpiling of Bacteriological (Biological) and Toxin Weapons and on Their Destruction," usually cited as the "Convention on Biological and Toxin Weapons."

When the Convention was signed, bioweaponry was practically useless; whence the treaty's brevity and apparent comprehensiveness. Widespread death to humans and livestock and great damage to crops and forests could readily be accomplished with microbes or microbial products. But uncontrollability and unpredictability were disqualifying tactical weaknesses, though they had not proved perfect discouragements to research and development (and would continue not to prove so under the Convention).[51] In 1969, the United States, the first and only "life-sciences hegemon," had announced that it would unilaterally give up on bioweaponry, a class of arms more valuable "on the table" than on the battlefield.

The Convention was typical of its class: an end in itself for the activists who sought it, a means to an end for the statesmen who signed it. Not even the scholarly imagination was captured. In his vast diplomatic history of the period, Raymond L. Garthoff has overlooked the Convention entirely, even for what it was, an appetizer chosen for the next month's entree, the first Nixon-Brezhnev summit meeting, at which was inaugurated an era of strategic-arms "limitation."[52]

"Progress" in politics is always revocable, and "détente" in its first incarnation did not survive the decade. Progress in *science,* though, is cumulative. Synthesis of the first molecule of recombinant deoxyribonucleic acid (rDNA) in 1972 changed permanently humanity's relationship both to its evolutionary past and to its technological future. It is likely that 1972 will be remembered more in the twenty-first and twenty-second centuries for directed genetic recombination than for the Moscow Summit. Whether or not—and in what sense—it is remembered for the Convention on Biological and Toxin Weapons is another matter.[53]

IX

PURSUING THE NATIONAL POLITICAL ADVANTAGE

DIFFERENTIATING between national security interests and national political interests is not always possible, as Clausewitz would have argued. The former might be considered an extreme form of the latter, or both might be identified on the same spectrum but at its opposite ends, the spectral parameter being some function of physical risk.

At the national-security-interests end of the spectrum would be cases of violent struggle—potential, impending, or immediate, and sometimes desperate. At the other end of the spectrum, the national-political-interests end, would be cases in which strategic problems are never immediate or desperate, cases in which humanitarian problems may openly be acknowledged and the moral discomfort of governments and individuals may even be admitted and discussed candidly. In these cases, nations pursue their interests less violently or less directly, though not always less destructively.

We will study as an example of decisions made at the national-political-interests end of this spectrum a case involving clinicians more than researchers, beneficence more than contracts, filth and fever and hunger more than high technology.

Marginalizing Life, Health, and Happiness

The foundation stone of the United Nations is "the principle of the sovereign equality of all its Members."[1] Accordingly, the voting equality of sovereign states is an elemental feature of the United Nations General Assembly. However, voting equality is not found in all United Nations bodies. In the United Nations Security Council, permanent members are joined by temporary members, but only permanent members can veto Council resolutions, and, in specialized agencies and commissions, full cooperation of large, rich, and technologically advanced states is often functionally indispensable, so "influence equality" does not prevail, foremost members enjoying more than their formal say. Yet, from time to time, several specialized bodies have come to be dominated by factions and administrations openly disdainful of leading members, especially the United States, despite the unequaled size of the American contribution to the overall United Nations budget.

Article 17 of the United Nations Charter stipulates that "[t]he General Assembly shall consider and approve the budget of the Organization [the United Nations]" and that "[t]he expenses of the Organization shall be borne by the Members as apportioned by the General Assembly."[2] Every joining state, including the world's most powerful, has accepted this stipulation, along with the Charter's Madisonian mechanism for the adoption of amendments.

Early on, the United States contributed funds equaling almost 40 percent of the United Nations budget. In 1954, the U.S. contribution was lowered to a third and, in 1972, at American insistence, to a quarter. Each of these figures, though impressively large, represented a budgetary share lower than America's contemporaneous share of the world's economic output, as well as a per capita budgetary share lower than that paid by the citizens of many smaller countries. However, in the mid-1980s, about half the membership still paid nothing or next to nothing.[3]

For some years after its foundation in 1945, the United Nations was friendly territory for the United States. But long before the 1980s, the United States had come to feel insulted, ignored, or impotent in the General Assembly, in the International Court of Justice, and in the various United Nations agencies whose individual budgets it underwrote but whose policies it could often not affect to its satisfaction.

This historic shift in international attitudes toward the United States had many partial explanations beyond jealousy of the weak for the strong: the perception of American support for dying colonial empires and for their post-mortem residua; the spread of Marxism, Leninism, Maoism, and authoritarian socialism; American meddling or armed intervention in areas where America's interests were illegitimate or where its legitimate interests were only marginally affected; the rise of the "Non-aligned Movement"; American support for the states of Israel and South Africa and for "anticommunist" regimes of whatever quality; the rise of Islamic fundamentalism; and commercial trade policies and environmental policies suggesting to many critics an American divine-right-to-wealth assumption.

American attitudes also shifted.[4] The representatives of the international community seated in the United Nations General Assembly and employed in its various administrations were seen more and more as hostile, petulant, incompetent, intransigent, corrupt, lazy, and, worst of all, ungrateful.

Many within the American government and electorate came to favor a policy whereby contributions assessed by the General Assembly would be withheld. Principled refusal to pay was not unprecedented; Nikita Khrushchev (1894–1971) had refused to pay the Soviet share of peacekeeping expenses incurred during the Congo crisis. But, in Washington, as late as 1978, the legal adviser of the Department of State had concluded that a policy of withholding would violate a formal treaty obligation and could under no circumstances be considered legal. Nevertheless, in 1980, selective withholding began on a small scale.

In 1983, Senator Nancy Landon Kassebaum of Kansas, hoping to end the "tyranny of the Third World majority in the UN,"[5] proposed an amendment to the State Department authorization bill—the so-called "Kassebaum Amendment"—requiring that assessed contributions to the United Nations (UN), the United Nations Educational, Scientific, and Cultural Organization (UNESCO), the World Health Organization (WHO), the Food and Agriculture Organization (FAO), and the International Labor Organization (ILO) be restricted in certain ways starting in 1984. In 1985, the House of Representatives easily approved a further restriction known as the "Solomon Amendment," and Senator Kassebaum reintroduced her earlier initiative with the new stipulation that full payment of future contributions be made contingent upon acceptance of contribution-weighted voting on all budgetary matters considered by the United Nations and its agencies. A reconciled "Kassebaum-Solomon Amendment" was approved by Congress and was signed into law by President Ronald Reagan. This new law reduced the American contribution to no more than 20 percent of the United Nations budget. At the same time, United States federal budgetary sequestrations mandated for 1986 by a feeble deficit-control measure, the Gram-Rudman-Hollings Act, further shrank the American contribution.[6] The net effect was a sharp and sudden drop in operating revenues for the United Nations system.

Technically, restoration of the Kassebaum-Solomon cuts was possible only on ratification of a Charter amendment abandoning the sovereign-equality-of-states principle. Amending the Charter in this way would surely have been impossible; amending it at all was highly impractical. In the event, the General Assembly offered the United States a "gentleman's agreement" to be more solicitous in future.[7] And the United States Government, criticized at home and abroad, seems eventually to have accepted the compromise. At any rate, Vice President George Bush, one-time ambassador to the United Nations and then a presidential candidate, said in August 1988 that the United States should pay its United Nations debt,[8] and President Reagan, in September, promised to pay assessments in arrears, citing recent United Nations successes in peacemaking.[9] Early in the Bush presidency, however, Secretary of State James Baker renewed the threat of a funding cut, this time to discourage the WHO from granting membership to the "state" not long before proclaimed by the Palestine Liberation Organization.[10]

In these manipulations, funds were denied not only to United Nations agencies and operations annoying to the United States but also to the World Health Organization, against which few complaints were then being lodged, at least in public. In February 1988, long after withholding had begun and well before it had moderated, the State Department, explaining its funding policies in materials furnished to Congress, cited the WHO among the "second priority cluster" of international organizations. Inclusion in this "second priority cluster" marked the WHO as "particularly effective, well-managed and of critical importance." The only higher "cluster" contained the North Atlantic Treaty Organization. The fourth and lowest contained the United Nations as a whole.[11] The WHO was so highly respected—and, being based in Geneva, so far from the East River—that it was popularly associated less with the United Nations than with its own projects, such as the worldwide eradication of smallpox and ongoing efforts to lower infant and childhood mortality rates in developing countries.

Nevertheless, at great risk to millions of the poorest, sickest, and least threatening people in the world, the Kassebaum-Solomon withholdings specifically included the WHO. At a time when Gram-Rudman-Hollings budgetary sequestrations were likely to occur. At a time when assessed payments were also being withheld by fifty-seven other nations, most notably bankrupt countries, like Brazil, Mexico, and Poland, the war-waging Islamic Republic of Iran, and, apparently for no good reason, Saudi Arabia. *And* at a time when the acquired immunodeficiency syndrome (AIDS) was fully realized to be pandemic.

To what effect?

By the end of 1986, the WHO faced a critical shortfall of funds requiring submission of a "phantom budget" and plans to cut back its activities by at least 10 percent. The director general, Halfdan Mahler, complained that the "W.H.O. is being unfairly victimized because it belongs to the United Nations system. . . . Without justifying the sweeping criticisms of the United Nations, I repeat that W.H.O. should at least not be subjected to them indiscriminately. . . . I underline the word indiscriminately. Emergency action is required in the short term and soul-searching in the longer term to make up for the dollars."[12]

By the end of 1987, the United States was the only industrialized nation still in arrears, and it owed the WHO more than $100,000,000. The WHO had cut its staff, closed offices, and canceled programs, including programs in nutrition, sanitation, and malaria control. Resources were being withdrawn from long-term projects in a frantic shift to the AIDS containment effort, still funded only on a voluntary basis. To the editor of the *Washington Post,* an official wrote that the WHO was "getting weaker each day the U.S. contribution continues to be withheld" and that "the failure of the United States to meet its financial obligations is being widely perceived as a lack of political commitment by the United States to the fundamental principle of better health for all."[13]

On January 1, 1988, in an editorial entitled "A New Threat to World Health" published in *Science,* the scholarly journal of the American Association for the Advancement of Science, a leading American life scientist and principal in the WHO's vaccine development efforts, Barry Bloom, made the following comments, between whose lines there was much to be read:

> WHO is suffering the worst crisis in its history, and ironically we are the cause. Without reason or notice the U.S. Government has unilaterally and arbitrarily refused to pay its assessment. We are $118 million in arrears for 1986–87. That represents a cut of almost 25% in an annual budget that has had zero real growth for 6 years. No organization can function with an unplanned reduction in budget of that magnitude, and drastic cuts in its program and skilled personnel are now being made.
>
> WHO was caught by the Kassebaum amendment directed at punishing the United Nations' system because "the U.N. and its specialized agencies, which are financed through assessed contributions of member states, have not paid sufficient attention in the development of their budgets to the views of the member governments who are major financial contributors to those budgets." For WHO this is tragic because the objection simply does not pertain. WHO has its own budget process independent of the U.N.'s, and the U.S. Government representatives have repeatedly acknowledged that its views have invariably been taken into account in the budgetary process. Curiously, the United States since 1981 has praised WHO's efficiency and voted in support of the WHO budget. Yet we do not pay our assessment.[14]

Curiously, indeed. Bloom realized, through his own inquiries at the State Department and in the Senate,[15] that the WHO had truly been "caught," and not by accident. The tone of the World Health Assembly had irritated many in Washington, and several of the policies the Assembly had recently adopted had been inconsistent with the international political interests of the United States as they were then being perceived. Insight into American displeasure can be gained by reviewing the president's reports to Congress on United States participation in the United Nations for 1986 and 1987.

As it told the Congress, the Reagan administration (1981–1989) did not want the World Health Assembly to "engage in debate over political issues outside WHO's technical area."[16] For instance, "[a]lthough there was no relevant item on the Assembly agenda, the U.S.S.R. and others introduced a resolution endorsing the UN International Year of Peace and calling for a cessation of the arms race."[17] The United States also objected when "[t]he Assembly took up the long-standing issue of 'Health Conditions of the Arab Population in the Occupied Arab Territories, including Palestine.'"[18] Similarly, the United States cast the only vote opposing adoption of "the Assembly's traditional resolution concerning health conditions in southern Africa" since it "still contained objectionable political judgments and had been worsened, in the U.S. view, by the insertion of requests for assistance to national liberation movements, which the United States has always opposed."[19]

As it had for years, the administration objected to the WHO's efforts to regulate the corporate marketing of processed infant feedings in the developing world:

> In a statement to the Assembly, U.S. Delegate Neil A. Boyer expressed regret that [a new] draft resolution tried to increase the number of commercial products in which WHO had regulatory interest. Also, he said, the resolution's request that maternity wards and hospitals not use free or subsidized supplies of breastmilk substitutes "seems to be an unnecessary folly. It appears to impose a new and expensive burden on the hospitals of developing countries, which can ill afford the supplies they need," he said. ". . . Why should the global WHO be trying to dictate how the maternity hospital in your local community carries out its business?" Noting that the United States had opposed the breastmilk substitutes code of 1981, Boyer said the U.S. delegation had no option than to oppose an Assembly resolution that sought to extend WHO's regulatory interests. The resolution was adopted by a vote of 92 in favor, 1 (U.S.) opposed, and 6 abstaining.[20]

On the matter of "health for all by the year 2000," the United States again found itself in lone opposition:

> Another resolution focused on the repercussions on health of the "worldwide economic crisis" and appealed to industrialized countries to increase their cooperation with developing nations. The U.S. Delegation objected to what it said were extraneous political elements in the resolution, one pertaining to the "New International Economic Order" and another to the policies of international banks, which the United States said are not the proper concern of the World Health Assembly. For these reasons, the United States called for a vote on the resolution, and it was adopted by a vote of 78 in favor, 1 (U.S.) opposed, and 15 abstaining, most of them Western European countries.[21]

The politics of AIDS was treated more elliptically. The 1986 report noted the director general's forecasts of AIDS-related resource requirements: "WHO would need approximately $200 million a year in new resources immediately, and anticipated a need of approximately $1.5

billion a year by 1990."[22] No mention of AIDS was made in a detailed review of the drastic effect that the Kassebaum-Solomon and Gram-Rudman-Hollings withholdings were having on the WHO.[23] A bit lamely, "[t]he U.S. Delegation pointed out that the financial difficulties were broad and not attributable to just one country."[24] In the president's 1987 report, the American delegation described the United States as "a major supporter of the WHO AIDS program," citing $5 million from U.S. Agency for International Development funds diverted to a WHO grant, as well as $9 million in technical support and $3 million worth of condoms for AIDS programs in the developing world.[25]

In general reference to the WHO's financial crisis, Donald Newman, under secretary of health and human services and head of the U.S. delegation, told the 1987 Assembly that "the problem faced by WHO in relation to U.S. assessments confronts much of the UN system. WHO has not been singled out for discriminatory treatment. We believe, in fact, that WHO is leading the way among the specialized agencies in seeking improvements in the budget process."[26] But he added that the WHO had to understand the need to economize and could not expect increased contributions when many governments already were not paying their past and current bills. Presumably to help the WHO understand this point even more vividly, "'As a symbolic gesture reflecting our concern over the need to reduce expenditures,' he said, the U.S. Government would decline its allocation for WHO program activity in the United States and ask that the overall budget be reduced by that amount. He invited other industrialized countries to consider taking similar action."[27] This "gesture" was made despite repeated protestations sent to and made in Washington by WHO officials detailing the "damage being done to WHO programs" by United States policies.[28]

Director General Mahler, who in 1987 withdrew his name from consideration for a fourth five-year term,[29] addressed the Assembly poignantly: "What crimes has WHO committed against those who are withholding mandatory contributions? . . . That it has stimulated member states to adopt health policies in line with WHO health culture? That it has saved them more than they have ever contributed to WHO by eradicating smallpox? That WHO has taken the international lead in the battle against AIDS. . . ? Or that your Organization has displayed outstanding fiscal responsibility?"[30]

In September, Mahler wrote to Otis Bowen, U.S. secretary of health and human services and a physician, reporting deep program cuts in some of the world's neediest countries. Two months later, he wrote again to revise his report; a shortfall in U.S. contributions even larger than anticipated had made additional cuts necessary. Insolvency was the only alternative.[31]

In response, the State Department's John C. Whitehead wrote the following: "I am very aware [of] and disturbed about our payment situation in the WHO and other international organizations. . . . I can assure you of this Administration's strong support for the fine work being done by the WHO and of your outstanding leadership. I also can assure you that we will continue to work vigorously in seeking to satisfy our financial obligations to the WHO, an organization for which the United States has great respect and appreciation."[32]

Further on, the president's 1987 report to Congress reviewed the year's progress toward the larger American political objective: "consensus based budgetary decision-making reforms" contained in resolution 41/213, the "gentleman's agreement" referred to above. In the General Assembly, American efforts had failed. But in the major specialized agencies, American pressure had been more effective. "Success was achieved in the WHO, ILO, UNIDO [United Nations Industrial Development Organization], WMO [World Meteorological Organization], and ICAO [International Civil Aviation Organization]. The FAO was the only specialized agency in which the United States tried, without success, to achieve an appropriate adaptation of the consensus based budgetary decision-making reforms."[33]

As far as the United States Government was concerned, the agonies of the WHO had served a higher purpose. "Success" had been "achieved."

There is no reason to think that anyone involved in this case at any time calculated that it was in the national interest of the United States for babies to be unprotected against polio or for mothers bitten by mad dogs to be denied access to properly refrigerated rabies vaccine. Nor is there any suggestion that the WHO, both in concept and in practice, was ever considered by the United States government to be anything other than a great invention—and a net political asset.

This is not to say that the eccentric ethical views sometimes attending international public-health controversies never found expression in government actions during the period discussed. It may well be that some people within policymaking circles sincerely believed that AIDS was the wrath of a just god or that grand multiparity was the blessing of heaven or that healthy people had no moral obligation to help sick people or that the rich actually had an obligation *not* to help the poor. Nevertheless, we will hereafter assume the prevalence of common secular ethical opinions among the individuals who made relevant decisions.

That acknowledgment aside, three analytical tracks present themselves.

Along the first track runs the notion that the WHO was unfair to the United States and that restitution of institutional justice in some way required or excused performing fiscal and administrative violence against the WHO or against the tyrannical "Third World majority" controlling it. This first track is deontological or absolutist. It terminates quickly in a dead end, since most American complainers and reformers were far more annoyed at the United Nations in general than at the WHO in particular.

Along the second track runs the notion that the WHO seriously needed to be reformed in the 1980s, and that the long-term universal benefit that would follow its reform would justify the high costs paid by those innocent masses who were then dependent upon the uninterrupted functioning of its projects. This second track is consequentialist and, specifically, utilitarian. It does not lead to much of interest in this case, though it does serve one observation: some of those opposed to international restrictions on corporate behavior in the developing world were ideologically committed free-traders, predisposed to utilitarian ethical reasoning. Their opposition, when ideologically based, could be classified as a "liberal enhancement" of the realist illusion.

Along the third track run most of the facts of this case, and their terminus is a non-Kantian, illiberal ethical error recognizable as a variant of the realist illusion, adapted here to the environment of international organizations.

The ethical reasoning displayed was non-Kantian—in a simplified sense—on three counts. The WHO's beneficiaries were being used as the "means" to achieve the "end" of WHO reform. The WHO itself and the WHO's beneficiaries were being used as "means" to achieve the "end" of UN reform. The UN and its constituent WHO and the WHO's beneficiaries were being used as "means" to achieve the "end" of advancing American political interests internationally.

The ethical reasoning displayed was illiberal since it was not *both* rights-based and individual-regarding. Nation-states have rights in the United Nations Charter, and, therefore, the ethical reasoning was in part rights-based. But it was not *human*-rights-based.

The ethical reasoning displayed was characteristic of the realist illusion in several respects. First, hostile actions were being taken in an unregulated environment in which treaty obligations could be set aside with minimal concern. Second, hostile actions were being directed against durable personified entities and their combinations—governments, nations, the "Third World," international organizations—and *not* against mortal *persons*, even though, intellectually, "hostility" toward persons in perilous circumstances could not have been overlooked as the final effect. Third, there was the ethical assumption on the American side that the actions of great states were important actions, that the rights of great states were important rights, that the voice of a great state, of a great contributor, should rightly have been heard over the voices of lesser, weaker, poorer states, however many individuals they contained, however great their collective need.

X

AGGRANDIZING THE CORPORATION

ECONOMIC corporations act in beneficial ways. They concentrate capital and labor. They underwrite product research and product development, and they bring useful goods efficiently to market. They create jobs and wealth, and they sometimes redistribute small amounts of the latter philanthropically. Representing the common interest, leaders of capitalist economies encourage corporate formation, corporate vitality, and corporate growth.

Appreciations offered where deserved, we now turn to finer points of conduct.

We have noted Hobbes's comment on the statelike nature of corporations: "lesser Common-wealths in the bowels of a greater." His was a feudal analogy presumably intended to suit the great trading companies of the West, but it need not have been parochial. The analogy would have suited just as snugly the corporations then forming in a truly feudal society, Tokugawa Japan, one of Hobbes's "dry places."[1] Nor did the analogy need specify as a corporation's residence the gut of a single "greater commonwealth." International trading companies were even then inventing colonialism, and they would rightly be cited several centuries later as having been the antecedents of great transnational and multinational corporations, some superior in assets to their smaller sovereign hosts.

Do corporations resemble the Leviathan? Structurally and functionally, some do and some do not. Have corporations found realist ethical arguments well suited to the defense of their more controversial acts? Some have, certainly, though large modern corporations sensitive to popular judgment have not often advertised their moral reasoning as having been inspired by Machiavelli or Hobbes.

A more presentable patron has been Adam Smith, the first political economist and a liberal. Smith is remembered popularly as having advocated free-market solutions to problems of trade and economic growth. In 1776, Smith stated his arguments in *An Inquiry into the Nature and Causes of the Wealth of Nations.* Under conditions of perfect liberty, eventually called laissez-faire capitalism, the general prosperity would be advanced through the pursuit of individual self-interest: "It is not from the benevolence of the butcher, the brewer, or the baker, that we expect our dinner, but from their regard to their own interest."[2] Furthermore, the imposition of rules on economic transactions would not only decrease the number of coins in the tradesman's purse; it would also decrease the amount of beef, beer, and bread on society's table.

Did the economic liberalism founded by Smith extend the political realism founded by Hobbes? Yes, in two ways.

First, it provided for economic realists a legitimate consequentialist argument against the imposition of rules on economic transactions—even benevolent rules and especially redistributive rules.

Second, it complemented Hobbes's pro-international-anarchy argument in a fashion Hobbes could not have managed himself. Hobbesian international anarchy was not an inevitability; it was a rhetorical weapon in the fight against the catholicism of the pope; it was a theoretical companion piece for the nationalization of religion. The benefits of Hobbesian international anarchy were *intra*corporate; the benefits were seen in domestic society. By contrast, the benefits of Smith's free trade were either *intra*corporate or *inter*corporate; the benefits were seen by any individual, any factory, any society, any nation that participated; each would enjoy an increase in welfare.

The Wealth of Nations and the free-trade tradition it founded made possible the "liberal enhancement" of the realist illusion, the dual claim that the amoral pursuit of self-interest is rational and that the generalized amoral pursuit of self-interest inevitably serves the general welfare.

"Amoral pursuit"? Smith would have objected to any association of *The Wealth of Nations* with amorality. Smith's fame first rested on *The Theory of Moral Sentiments* of 1759; the great work that eclipsed it in 1776 was in many ways its spiritual sequel. Much of the earlier book concerned the modulation of self-love, not its indulgence. For instance:

> When the happiness or misery of others depends in any respect upon our conduct, we dare not, as self-love might suggest to us, prefer the interest of one to that of the many. The man within [the conscience] immediately calls to us, that we value ourselves too much and other people too little. . . .
>
> One individual must never prefer himself so much even to any other individual, as to hurt or injure that other, in order to benefit himself, though the benefit to the one should be much greater than the hurt or injury to the other. . . .[3]
>
> Those [these and other] general rules of conduct, when they have been fixed in our mind by habitual reflection, are of great use in correcting the misrepresentations of self-love concerning what is fit and proper to be done in our particular situation.[4]

Smith thought men and women acted rationally and, for the most part, in their own interests, but he was not an egoist like Hobbes, whose "doctrine . . . was . . . offensive to all sound moralists."[5] Smith's advocacy of *free* trade was not an apology for *unfair* trade.

Less timidly, Milton Friedman, though famous as a libertarian and not as a contractarian, has advocated for corporations an ethic close to Hobbes—so close that few executives would feel safe endorsing it outside a boardroom or a closed meeting of major stockholders. Adapting lectures delivered six years earlier, Friedman in 1962 produced a short book entitled *Capitalism and Freedom.* Therein he wrote, "Few trends could so thoroughly undermine the very foundations of our free society as the acceptance by corporate officials of a social responsibility other than to make as much money for their stockholders as possible. This is a fundamentally subversive doctrine."[6] Here was the drama of Hobbesian international ethical theory—self-help the only salvation under anarchical conditions—being played *domestically,* the corporation taking the part of the state and altruistic "social responsibility" taking the villainous role of papal transnationalism. Here was a libertarian's adaptation of the "best" of Hobbes.

Embellishing his argument, Friedman published in 1970 in *The New York Times Magazine* a widely noticed article introduced as "A Friedman Doctrine": "The Social Responsibility of Business Is to Increase Its Profits." He criticized businessmen who claimed

> that business is not concerned "merely" with profit but also with promoting desirable "social" ends; that business has a "social conscience" and takes seriously its responsibilities for providing employment, eliminating discrimination, avoiding pollution and whatever else may be the catchwords of the contemporary crop of reformers. . . . [T]hey are . . . preaching pure and unadulterated socialism. Businessmen who talk this way are unwitting puppets of the intellectual forces that have been undermining the basis of a free society these past decades. . . .
>
> A corporation is an artificial person and in this sense may have artificial responsibilities, but "business" as a whole cannot be said to have responsibilities, even in this vague sense. . . .
>
> In a free-enterprise, private-property system, a corporate executive is an employe of the owners of the business. He has direct responsibility to his employers. That responsibility is to conduct the business in accordance with their desires, which generally will be to make as much money as possible while conforming to the basic rules of the society, both those embodied in law and those embodied in ethical custom. . . .
>
> Of course, the corporate executive is also a person in his own right. As a person, he may have many other responsibilities that he recognizes or assumes voluntarily—to his family, his conscience, his feelings of charity, his church, his clubs, his city, his country. . . . But in these respects he is acting as a principal, not an agent. . . .
>
> [T]he doctrine of "social responsibility" involves the acceptance of the

> socialist view that political mechanisms, not market mechanisms, are the appropriate way to determine the allocation of scarce resources to alternative uses. . . .
>
> [P]recisely the same argument applies to the newer phenomenon of calling upon stockholders to require corporations to exercise social responsibility. . . .
>
> [T]he use of the cloak of social responsibility, and the nonsense spoken in its name by influential and prestigious businessmen, does clearly harm the foundations of a free society.[7]

Here were minor Leviathans, self-determined in a nearly natural state. Friedman did acknowledge that "artificial persons"—"lesser Commonwealths in the bowels of a greater"—might have "artificial responsibilities," but these responsibilities were at best "vague," and they *should* not redirect the pursuit of corporate self-interest.

Here, too, was Hobbes's theory of obligation, this time applied not to citizen and "Soveraign" but to employee and corporation. The corporation was no less an artificial person than the nation-state and was barely less jealous. Note that Friedman took pains to eulogize his corporately amoral employee-executive as a thoroughly decent and charitable human being *only in his ethically evaluable life,* his *non*corporate life.

There was also in this "doctrine" another hint of Hobbes—and here "liberalism," as in free-market economics, extended the reach of realism. Hobbes argued *toward* international anarchy on moral grounds. Friedman argued *toward* the social nonresponsibility of corporations also on moral grounds. The economic-welfare-enhancing effect of markets was optimized when markets were unrestrained by politics, when capital accumulation was undiminished by redistribution.

Free-market romanticism is one thing; common corporate practice may be another. Of the latter, what can we say, and, to be fair, what must we be able to say?

The frequency with which corporate executives follow the "Friedman Doctrine" in planning or defending their actions is unknown, though not unknowable. Altruism still lurks in the least likely places, despite Friedman's well-intentioned efforts to restrict its range. Yet, even more surely, in the corporation as in the state, self-interest thrives unashamed and unthreatened.

But frequency is not really the issue here. Mechanism and potential effect—*these* are the issues. Corporations, like states, do act in ethically illiberal ways, all the while directed by men and women who "may have many other responsibilities that [they recognize or assume] voluntarily—to . . . family, . . . conscience, . . . feelings of charity," and so forth.

For all the similarities between states and corporations, the former are and the latter are not sovereign, in the common sense. This is a critical difference politically and legally. Is it a critical difference ethically? If it were, then one would expect the ethical defense of state action to differ from the ethical defense of corporate action primarily on grounds of sovereignty. States might automatically be excused for conventionally immoral actions because they were sovereign, while corporations might be condemned for similar actions because they were not sovereign, not immune to destruction by government. Conversely, states might be taken more harshly to task than corporations because only states, as sovereign entities, could enforce moral standards, while corporations might be held less culpable because they committed only those sins allowed or encouraged by their overseeing states. So, sovereignty alone seems an indeterminate ethical criterion.

Can the ethical impact of sovereignty be better explained not by sovereignty *per se* but by associated qualities? Grandeur, nobility, immortality, divinity? The answer here might be yes. Soldiers routinely give their lives for comrades and country; employees may give their free time for coworkers and corporation, but nothing much beyond that. For two reasons, perhaps. Exaggerated sacrifice for some social object less than the state competes with and therefore degrades the sacrifice offered to the higher object, the state itself. And, more directly, states are perceived as transcending individual life. They contain an individual's family, and they protect and enhance his or her genetic, ethnic, and racial future. States are seen as permanent, as intrinsically worthy, and often as divinely endowed or divinely allied. Corporations are rarely seen in these ways.

Accordingly, the policies of corporations are less likely than those of states to be identified with high moral principle and the common good. The policies of corporations—again, successful ones—are less likely to be perverted by passion, more likely to be coldly rational. Oddly, this makes corporations, more than states, perfect realist creatures, more readily observed within the realist frame, more easily seen as the objects of realist theoretical modeling. But it also gives them less access to the indulgence of realist ethics. Corporations can be seen as group-persons; legally, corporations are, in fact, seen as and sued as group-persons; theoretically, they can even be seen as "structurally restrained" moral actors, so well designed to pursue profits as to be incapable of pursuing moral ends.[8] But corporations are not credible beneficiaries of ultimate sacrifice. Their survival and prosperity cannot be overriding moral goods, not ends in themselves. When we consider corporations as recipients and perpetuators of the realist ethical tradition, we do so to understand them as ethical actors, not to endorse them as Hobbesian "Mortall Gods."

Yet, from time to time, corporations have grown so great as to become quasi-states, the old East India Company being the most convincing example. More recently, in the period of Western and East Asian industrialization and in the postcolonial period, many large corporations have behaved in ways disrespectful of governments and societies—both "home" and "host" governments and societies. Many corporations have fouled their environments. Many have produced or sold dangerous goods. Some have interfered in political processes through ordinary corrupt practices, such as bribery, others through acts of intrigue and subversion.

It is not our task to recount these instances, even in general; they are too numerous, too intricate, and too often the result of simple immoralities, such as greed or lust for dominance. What we will be considering in this section is more abstruse: the ethical arguments by which corporations and their advocates have defended actions known to be harmful to human health or well-being.

How can we fairly describe the ethical reasoning that has led to a particular corporate policy or guided the formulation of its defense? Provocative corporate actions are rarely taken by scholars of known philosophical persuasion or by their confessed adherents, and public statements of corporate philosophy and intention are typically tendentious, if not dishonest. Our ability reliably to detect ethical reasoning in the acts of corporations is modest indeed. And even when reliable detection is possible, no inference can be drawn about the prevalence of one type of reasoning or another in the broader corporate world.

What we *can* do is this. We can set out a range of illiberal corporate behaviors and then assay for realist content corporate actions falling within that range. We wish to consider the effects of a substitution of realist ethics for liberal ethics in matters involving the life sciences. We can manage well enough without formally assessing behavioral prevalence.

At one end of this range are illiberal behaviors occurring when corporate welfare has *marginally displaced* "individual" welfare as first priority. These behaviors imply that "individual" welfare has not been abandoned entirely as a goal but has come to be served along with, only if, or only after corporate welfare has been served.

Not included, even at this end, are corporate decisions to forgo nonprofit-making socially beneficial activities. Genentech's reluctance to undertake research on a malaria vaccine, even while being urged to do so by the World Health Organization,[9] seems to have been explainable as risk aversion. Corporate welfare was not displacing "individual" welfare, the welfare of individuals singly or collectively at risk for malaria. Corporate welfare and "individual" welfare were just following nonparallel paths. If health-policy decision-makers at national or international levels had expected private involvement in a financially dangerous vaccine program, then adequate public guarantees should have been forthcoming—as they often are forthcoming in similar circumstances. In the United States, the National Institutes of Health (NIH) funds much socially beneficial research performed by investigators who are employees of private corporations. In the malaria case, Genentech decided that it was unable to help, unable to take on what was actually an underfunded public responsibility. True, Genentech acted in accordance with the neo-Hobbesian "Friedman Doctrine," but only in commonsensical accordance with a nonmalignant version of that doctrine. Genentech was not pushing its corporate advantage to the limit of the law; it was not actively doing harm; it was not explaining the infliction of suffering as the market price of liberty.

Examples of "marginal displacement" are common in the corporate practice of modern clinical medicine in the United States. Many American physicians have ceased to be independent economic agents by becoming employees of growth-conscious corporations: private hospitals, health-maintenance organizations, free-standing clinics and laboratories, and the like. In this corporate mode, some physicians have found their clinical judgment strongly influenced by their employer's economic ambitions. A passage from a widely read anecdote—the setting was a for-profit "walk-in" clinic—suggests one of many mechanisms:

> The message doctors were given may have been that to work one's way up the corporate ladder, one had to work one's way up the patient-pricing ladder. The message that one should avoid being at the bottom of the ladder was considerably less subtle. The three full-time doctors who were at the bottom of the list a year ago are now all gone. One was told he would be fired if his charges did not increase. He resigned. Two were fired. Of these two, one was given absolutely no explanation, verbal or written. I am the other. I was told directly (but never in writing) by the corporate president that I was being fired for not having charged enough per patient.
>
> I am neither absolutist nor doctrinaire in avoiding laboratory tests. Like all clinicians, I order my share of x-ray films that I know will be negative, of cultures whose outcome will not necessarily change my treatment, and so forth. . . . By a standard of absolute necessity, I order too much. By the corporation's financial formulas, of course, I did not order enough. . . .
>
> In some ways, it is difficult to find fault with a corporation's focusing on the bottom line to increase profits. Such is the nature of business. Such is also the crux of the problem. The ethos of medicine and the nature of its primary concern—human health and emotions—make medicine a commodity less amenable to harsh business realities than other economic goods such as automobiles, hair spray, or lumber.[10]

We should note that susceptibility to monetary temptation has been a fixture of the physician's character for centuries, long before corporatization, and that the corporate president described above as greedy was himself a physician. We should also remind ourselves that this example of "marginal displacement" illustrates a mechanism whose political-ethical inspiration, if any, cannot be known to us. Formal realist thinking is not documented here, though something like the "Friedman Doctrine" does seem to have been observed. Disregarding the matter of inspiration, some have offered a Marxist explanation,[11] in all seriousness and rather skillfully.

"Wide displacement"—at the midposition on our severity spectrum—is also easily illustrated. Here, corporate action is overtly irresponsible, even frankly unethical, but still somehow "within the rules." Not too outrageous to be explained as incidental to the achievement of a greater good. And roughly inside the bounds of home-country or host-country civil traditions and statutes.

Some cases exhibiting "wide displacement" have involved multinational corporations behaving differently at home and abroad. The conduct of large pharmaceutical firms in the developing world has repeatedly been cited.

In poorly regulated markets, multinational firms such as Ciba-Geigy and Hoechst promoted for the treatment of common fevers, tension headaches, and pains of similar seriousness an "over-the-counter" oral antipyretic analgesic—the variously named and imaginatively compounded agent, dipyrone—banned for years in many countries after being linked convincingly to agranulocytosis, a form of aplastic anemia or bone marrow failure ordinarily lethal without sophisticated medical management, including intensive blood-component transfusion support.[12] Similarly, many firms inappropriately marketed the broad-spectrum antibiotic, chloramphenicol, encouraging its use for trivial indications long disapproved by the United States Food and Drug Administration; Parke-Davis, a major early offender with its "Chloromycetin" brand, in time became an industry leader in responsible labeling.[13] Chloramphenicol, unlike dipyrone, is an important agent; it has saved many lives. But, *like* dipyrone, it sometimes causes bone marrow failure. Chloramphenicol has been grossly overused in countries where it has been aggressively marketed; in Mexico, for instance, chloramphenicol has been sold briskly "over the counter." While hard to miss in circumstances of medical abundance, bone marrow failure—presenting as fatigue, pallor, or hemorrhage, as fever or overwhelming infection, as uncharacterized terminal illness, or, simply, as unexplained death—must often have remained diagnostically obscure in circumstances of public-health chaos.

Winthrop, Ciba-Geigy, and Schering promoted the use of anabolic steroids in African children; malnutrition was one advertised indication, "excessive fatiguability" another. Winthrop also marketed "Winstrol," its brand of stanazolol, a synthetic testosterone analogue, in Asia and Latin America. In Sri Lanka, Winstrol was recommended as having "demonstrated beneficial effects" when used as an aid to growth in sickly children aged one-and-a-half to six-and-a-half years. To facilitate compliance, Winstrol was made available in a pediatric syrup.[14] Compared to bone marrow failure, the masculinization of young girls must have been easily recognized, although its favored explanation and management may not in all communities have been scientific ones. In both boys and girls, premature closure of the growth plates in long bones would have been harder to spot, especially in children whose small size had prompted their anabolic steroid "therapy" in the first place.

Corporate abuses such as these sit at the midposition on our severity spectrum according to several criteria. Overlooking the high-risk–low-benefit therapeutic index of the obsolete agent, dipyrone, the products ill-applied here could have been manufactured and marketed by workers with the best intentions. The sale of these products was not against local law. Product information supplied to local physicians and to customers probably met or exceeded local standards. And, when host governments did finally object, as happened in Sri Lanka,[15] conduct sometimes improved.[16]

Let us now consider problems at the far end of our severity scale, where corporate welfare has *totally displaced* "individual" welfare as a priority, where "individual" welfare has effectively been removed from a corporation's operative list of concerns. Examples of "total displacement" are easily found in the histories of corrupt, rapacious, or irresponsible corporations. But they are also found in the histories of highly regarded firms.

The Johns Manville Corporation long suppressed data demonstrating the disastrous health effects of exposure to asbestos fibers and to its own asbestos-containing products.[17] It continued to sell these products for use in hospitals, schools, and elsewhere. It also continued high-risk asbestos manufacturing in Brazil for seven years after similar operations had been shut down in the United States.[18] Johns Manville was not the only asbestos offender, but it was by far the largest. In 1989, the cost of asbestos abatement and removal in the United States over the decade of the 1990s was projected to be from $100 billion to $150 billion,[19] figures whose height and imprecision support the guess that asbestos-related diseases, including the almost always fatal malignancy, mesothelioma, will be judged too costly to prevent.

In the 1970s, the Ford Motor Company projected the cost of making safe the explosion-prone gasoline tanks built into its hastily designed "Pinto" automobiles and some of its light-weight trucks. Fixing the problem would have cost at least $11 per vehicle. Ford compared the projected cost of correcting "all" defective vehicles—$137,000,000—with the legal fees and litigation awards likely to result from fatal and nonfatal burns sustained in uncorrected vehicles—a sum of $49,150,000. It figured that the corporate costs of correction would be $87,850,000 higher than the "social benefits"[20]—really the alternative corporate costs, *not* the social benefits. This calculation was crude and, apparently, informal. It must have anticipated the *industry-wide* cost of a mandatory product recall, since Ford's own production alone would not have explained the estimate.[21] Whatever its numbers really were or should have been, Ford chose to avoid both a recall of its products and a public warning to its customers and prepared instead to deny in court that Pintos and Ford trucks were less safe than comparable vehicles.[22] Ford's larger and wealthier rival, General Motors, may have made a similar calculation a decade later.[23]

The behavior of the Chisso Corporation was a particularly interesting example of "total displacement." In 1932, Chisso—the name means "nitrogen"—began producing acetaldehyde at its plant at Minamata, a fishing village on the Shiranui Inland Sea in Kumamoto Prefecture on the western side of Kyushu, the southern-most of the four largest Japanese islands. For seven years, Chisso had been paying a pollution indemnity to local fishermen, and community relations were generally good. The acetaldehyde production technique introduced at the Minamata plant in 1932 did require the use of mercury, a toxic substance, but small amounts of inorganic mercury diluted in Minamata Bay seemed not a great risk. And the plant was a major employer.[24]

In 1950, local sea life began to act abnormally; two years later, birds and cats, a year after that, dogs and pigs. Within a few more years, there had been so many "cat suicides" that several districts contained almost no cats at all. In April 1956, a 5-11/12 year-old girl presented to the pediatrics department at Chisso Corporation's factory hospital. Within five weeks, five more members of her family presented similarly with signs of a progressive peripheral and central neuropathy of unknown etiology. The hospital's physician-in-chief, Hosokawa Hajime, reported the outbreak to local public health authorities. Many more cases began to appear. By October 1956 "Minamata disease" had been recognized as heavy-metal poisoning and had been linked to the ingestion of local seafood.

In September 1958, ignoring the warnings of Dr. Hosokawa, its own employee, Chisso diverted its effluent to a new site. "Minamata disease" spread as Hosokawa had been predicting it would. By July 1959, mercury had been linked specifically to "Minamata disease." The livers, kidneys, and brains of dead victims were mercury-laden, as was the hair of living victims. It would later be proved that inorganic mercury was being incorporated into an organic molecule, methyl mercury chloride, within the Chisso factory itself, long before being concentrated in the food chain.

In October 1959, Dr. Hosokawa demonstrated conclusively to his employers that Chisso's own effluent produced "Minamata disease" in cats; Hosokawa's cat #400 would later become famous as the key experimental animal in this private in-house demonstration.[25] After cat #400, whose body was to disappear on its way to Tokyo University, Dr. Hosokawa and his staff were denied further access to acetaldehyde effluent. Fearing for the careers of the junior Chisso physicians who had assisted him, Dr. Hosokawa took full responsibility for having conducted such incriminating work. In November, Chisso ordered Dr. Hosokawa to suspend his experiments. His data were suppressed. When his experiments were allowed to resume in 1960, only effluent samples sent to him by the plant manager were tested. Dr. Hosokawa remained silent till 1969, when he testified against Chisso on his deathbed.[26]

From 1959 to 1973, there prevailed a legal and social controversy of extraordinary bitterness and paroxysmal violence. Despite its certain knowledge that it was causing profound neurological deficits in fetuses and newborns, in nursing infants, in children, and in adults and that it was bringing many members of its own community prematurely to the autopsy table—and that it was in less dramatic ways medically and psychologically and economically crippling many thousands more—Chisso continued secretly to discharge methyl mercury chloride until 1968, when the mercurial acetaldehyde production system became obsolete.[27]

At Chisso's end of the severity scale, where "individual" welfare is totally displaced by corporate welfare, corporations act as if they were waging mortal battle on a field of moral anarchy. More or less. Chisso was constrained a little by statute and by tort, as were Johns Manville and Ford. But evidence that Chisso felt *ethically* constrained is thin indeed. Perhaps it felt ethically *compelled* by an obligation to its stockholders or even to its employees, who surely needed their jobs but who, for the most part, did not avoid the local diet. Or, simply, Chisso may have felt ethically compelled by an obligation to itself; Japanese corporations have traditionally been objects of intense inwardly directed loyalty. For whatever reason or reasons, "social responsibility"—to use the "Friedman Doctrine" term—must have been judged an improper corporate attitude in Minamata.

Having described three regions on a spectrum of illiberal corporate behavior, we will now look closely at an intermediate case, a case of moderate severity, one of whose facets we have already examined in an indirect light.

Making the Very Best of an Illegitimate Market

It has always been true that some women have been unable or unwilling to nurse their babies. In premodern times, many women died postpartum; in premodern places, many still do. Other women for other reasons—personal, social, professional, educational, conjugal, and, rarely, medical—choose not to nurse their babies.

Historically, human newborns and infants denied access to their own mothers' milk have either died or been nursed by the mothers of other babies—by wet nurses—or by the mothers of nonhuman mammals. Foundling homes and hospitals often maintained lactating goats or donkeys specifically for the purpose of suckling babies, especially congenitally syphilitic babies, for whom wet nurses could not readily be secured. In some charitable institutions, "dry nursing"—administration of a grain-and-water pap—was routine. For species-specific reasons, unaltered cow's milk, an acceptable nutrient in older infants and children, could not safely be used as a predominant feeding in newborns or very young infants.

Before the modern era, the mortality rate among all babies, however fed, was similar to that observed today in much of the developing world. The mortality rate among babies wet-nursed was apparently higher than average, that among babies goat-nursed probably higher still. But the mortality rate among babies "dry-nursed" was often extraordinary; at a Dublin foundling home using this method exclusively, 99.6 percent of all babies taken in from 1775 until 1796 died—10,227 babies in all.[28] (Jean-Jacques Rousseau, *philosophe,* maternal-breast-feeding advocate,[29] and political ethicist, delivered to the Foundling Hospital in Paris all five children born to him by his illiterate hotel servant and wife-equivalent, the unhappy Thérèse la Vasseur.[30] The mortality rate for Paris foundlings, based on the years 1773–1777, was 80 percent.[31] Nobility, it seems, was for savages.) The quantity and nutritional characteristics of available feedings were surely important factors in survival. But most deaths, we may assume, were functions of immunological and sanitary factors.

A wet nurse did not produce colostrum for every new customer; colostrum, an immunologically protective milk component, was elaborated by the breast only in the first few postpartum days. The milk received from nonhuman udders must also have lacked colostrum, certainly human colostrum, and was suboptimal in other ways as well. But, still, goat-nursed babies did relatively well—assuming they sucked directly from the udder. One French physician recognized the mortality-risk differential between foundlings suckled directly at the human nipple or goat udder and those fed with milk expressed first into a collecting vessel. He postulated that milk exposed to air lost its "vital principle,"[32] that its nutritional value dissipated quickly. A more likely explanation by far was microbial contamination.

The most extreme dangers confronted newborns and young infants who were "dry nursed"—given a dry nutrient, such as grain, *in combination with water.* Babies fed in this manner constantly ingested water-borne infectious agents never before "seen" by their immunocompetent cells. Sooner or later, of course, these same infectious agents would be ingested, but the mortality risk in infections occurring later in infancy or in childhood was much lower.

Beginning in the nineteenth century, supplemental and substitute formulas became increasingly well devised. By the early twentieth century, sound recommendations and safe products were available for use in otherwise healthy full-term babies cared for fastidiously by literate adults properly instructed and supplied. Non-human-milk feedings typically were formulated from whole or skimmed pasteurized cow's milk plus extra sugar plus sterile water, or they were reconstituted from a premixed powder by the addition of sterile water. In the latter case, processed and augmented cow's milk remained the standard, though soybean-based preparations later came widely into use as well. By the late twentieth century, sophisticated nutritional research had led to the development of non-human-milk feedings designed specifically for critically or chronically ill babies unable to nurse or unable to tolerate human milk.

The development of artificial feedings for newborns and infants was a scientific triumph. And it was also a social revelation, particularly for well-to-do mothers who preferred to engage their domestic help on criteria other than lactation status. But modern mothers for whom bottle-feeding was the *less* convenient alternative were also strongly attracted to the practice, often because the new, scientifically designed formulas were thought superior to their "bestial" alternative.

Though the gross composition of human breast milk had at last been discovered, its finer distinctions and physiological benefits were not well appreciated. Breast milk had always looked a little thin to new mothers, many of whom had expected a fluid closer to the hearty whiteness of cow's milk. Doubts arose as to the vitamin and iron content of breast milk. Colostrum was still largely a mystery; it was sometimes discarded. And, objectively, formula-fed babies grew fatter than breast-fed babies. Science was improving on nature everyday; perhaps breast milk would soon be considered an evolutionary relic, like axillary hair and the appendix.

Or so thought many progressive-minded women and men through the first half of the twentieth century. And not just in rich, well-read, and well-watered nations but every place on the globe where artificial feedings were commercially promoted.

And promoted they were. By the second half of the twentieth century, while the advantages of breast milk were becoming less and less cryptic in the industrialized world, formula promotion in the nonindustrialized world was accelerating. Western-style advertising was used with great effect: posters, fliers, radio announcements, hut-to-hut solicitations by formula-company employees dressed as nurses. Local physicians were given money and gifts and local medical societies subsidized. Hospitals were given free feedings and obstetrical-ward supplies. Newly delivered mothers were given free formula samples, enough to last till lactation failed for lack of nipple stimulation. In targeted markets, breast-feeding rates fell drastically.[33]

Faith in technology has never required scientific sophistication; actually, faith in technology often seems inversely proportional to scientific sophistication. If magic milk from Switzerland, say, could get one's baby out of infancy strong and fat in a region accustomed to losing a tenth of its offspring in the first year, then it was worth buying. The irony, of course, was this: breast feeding could work well even under conditions of privation and poor sanitation, while artificial feedings presented unnecessary dangers under exactly these same conditions.

As the promotion of artificial feedings advanced in the developing world, the international health and development communities began worrying. Fewer babies protected by colostrum might mean more babies dying of orally acquired infections in the first few months of life. More babies fed formulas reconstituted with nonsterile water might mean more babies dying of orally acquired infections throughout the first year of life. More babies with diarrhea would mean more diarrhea in the general population. More diarrhea in the general population would mean more malabsorption and, therefore, more malnutrition and malnutrition-associated diseases, such as beriberi and vitamin-A-deficiency blindness. Fewer mothers actively nursing might mean more mothers conceiving again soon after delivery. Shorter child-spacing intervals might mean greater population pressure as well as more mothers and more babies suffering seriously from the protean effects of iron deficiency. More reliance on commercially supplied feedings in semi-barter or depressed economies might mean more babies presenting in hypoosmolar coma induced by the feeding of overly diluted—"stretched"—formula. More reliance on expensive commercially supplied feedings in poor societies might mean more diversion of capital resources away from savings and constructive enterprises.

In the early 1970s, as concerns became predictions and then as predictions began proving roughly to have been accurate, a ferocious controversy arose. It centered on the dominant corporate actor in transnational infant feeding, Nestlé of Switzerland, and it produced an international boycott, a famous vote in the World Health Assembly, excoriating recriminations all around, self-defeatingly extreme anti-formula attitudes among frustrated professionals in the field, and an extensive literature noted for its selective unreliability. A full recounting here is neither possible nor desirable.

Our interest in this case can be satisfied by considering three questions.

1. *Did contemporary professional opinion and data—nutritional, immunological, epidemiological—argue for or against aggressive marketing of artificial feeding products to newly delivered mothers in the developing world?*

Clearly, the great weight of professional opinion argued against such marketing. However, much of the evidence available in the early 1970s was anecdotal, or it applied to outmoded products; most of it had not been obtained under the developing-country conditions now at issue, and much of it was statistically assailable.[34] Eventually, however, fresh and reliable data confirmed the widely held opinion that artificial feedings were tragically inferior to breast feedings in the setting of poor sanitation.[35]

This is not to say that all anti-formula policies had been vindicated. As late as the mid-1980s, the United Nations Border Relief Operation (UNBRO), overseeing Cambodian refugee camps on the Thai-Kampuchean border, was still hostile even to the paraphernalia of artificial feedings: bottles and rubber nipples. Under UNBRO supervision, premature babies for whom breast-milk feedings could not be sustained were being spoon-fed. Bottles and nipples were not allowed. These babies regularly aspirated feedings into their airways and died.

Arguments against aggressive marketing were initially opposable on grounds of data insufficiency. And unreasoning hostility toward all artificial feeding methods remained just that—unreasoning. Nonetheless, during the period of interest, the 1970s and 1980s, the best available scientific opinion and the best available data did suggest—and rightly, as it turned out—that aggressive marketing did increase death rates substantially.

2. *How did Nestlé, taken as the prototypical marketer, see its own behavior, its own rights, privileges, and responsibilities?*

This is a difficult question. Nestlé, feeder of babies and Swiss chocolatier to the world, was accustomed to benign international regard. It took sharp offense at the brutal criticisms laid at its corporate door and went to great lengths to restore credit to its name, counterproductively on the whole. Was Nestlé a self-considered Leviathan waging economic warfare under conditions of moral anarchy? Did Nestlé think of itself as something of a "sovereign state,"[36] as has been said of another much-maligned corporation, International Telephone and Telegraph?[37] We do have an intriguing statement from an earlier Nestlé director: "We cannot be considered either as pure Swiss, or as purely multinational, i.e. belonging to the world at large, if such a thing does exist at all. We are probably something in between, a breed on our own. In one word we have the particular Nestlé citizenship."[38] Yet, this pronouncement, much quoted, was well within the bounds of oratorical flourish and preceded by several years[39] the controversy of interest.

More suggestive was the 1978 United States Senate testimony of Oswaldo Ballarin, Ph.D., chairman and president of Nestlé-Brazil. Senator Edward M. Kennedy's Subcommittee on Health and Scientific Research of the Committee on Human Resources was holding hearings on the infant formula problem. The U.S. Nestlé Company, which did not manufacture or market infant formula, had arranged for Dr. Ballarin to testify:

> [C]ertain groups are boycotting the sale of U.S. Nestlé products, such as coffee and chocolate. The boycott is for the avowed purpose of putting pressure on Nestlé's Swiss parent company to stop alleged misconduct in the marketing of infant formula in the Third World.
>
> I am aware of the specific charges made by these groups, and can state, based on my personal experience in many developing countries, that they are quite misleading and inaccurate.
>
> The U.S. Nestlé Co. has advised me that their research indicates this is actually an indirect attack on the free world's economic system. A worldwide church organization, with the stated purpose of undermining the free enterprise system, is in the forefront of this activity. . . . [T]he boycott . . . is made against one company that does not manufacture these products in the United States . . . , and . . . the film which has been distributed, under the title of "Bottle Babies," to many local churches and schools . . . has hurt, really, those who fight in a world of free enterprise.[40]

Here was a faintly Hobbesian scene, though faintly Hobbesian surely by coincidence and not by disciplined conviction. In *Leviathan,* Hobbes maneuvered to derive international anarchy as a supposedly immutable political-ethical principle—a fact and a good. He then based upon it (and upon other principles) a drastic and repressive corrective for domestic unhappiness, whose high pitch in Civil War England he thought chiefly the product of transnational religious meddling. In his Senate testimony, Dr. Ballarin portrayed "those who fight in a world of free enterprise" as being unjustly attacked by transnationalists—by "a worldwide church organization"—intent on causing economic loss and political conflict where neither otherwise would occur.

In the end, though, we can know much less clearly how Nestlé saw itself in the political order than we can know what Nestlé did in the societies served by its products. On grounds of action, if not intent, Nestlé and its lesser competitors did evidently displace "individual" welfare with corporate welfare.

3. *Upon what philosophical base did Nestlé and prominent defenders of full corporate prerogative rest their arguments?*

Here, we can say more. Several spokesmen for corporate prerogative wrote passionately about this case, and Nestlé is known to have endorsed their arguments and is also known, through a "leak" to the *Washington Post,* to have underwritten their efforts financially.[41]

In June 1980, in *Fortune* magazine, Herman Nickel, one of *Fortune*'s Washington editors—and soon to be United States ambassador to South Africa—published a polemical article entitled "The Corporation Haters." The Nestlé problem was Nickel's focus. Nickel castigated the "New Left," its "liberation" theologians, and its anti-business boycotters. Optimistically, he predicted, a strengthening-through-ordeal might be the long-term outcome of current attacks on major corporations: "Thus the Marxists marching under the banner of Christ may help the private enterprise system to adapt and survive—even though that may be the last thing they want to see happen."[42] The Marxists-marching-under-the-banner-of-Christ line had not been Nickel's own locution but the contribution of Richard Armstrong, one of *Fortune*'s two executive editors.[43]

The *Washington Post* revealed several important details about this article. In the fall of 1979, Nickel had been prompted to pursue the infant formula issue by Ernest W. Lefever, Ph.D., Protestant clergyman, Christian ethicist, international relations theorist, founding director of the Ethics and Public Policy Center at Georgetown University, and, soon enough, presidential transition advisor to Ronald Reagan. The Ethics and Public Policy Center offered to pay Nickel a $5,000 honorarium for the article. An infant-formula industry representative was to review the first draft.

After its appearance the following June, "The Corporation Haters" was recirculated by Lefever under the title "Crusade Against the Corporation: Churches and the Nestlé Boycott." All this occurred around the time that Nestlé gave to the Ethics and Public Policy Center donations of $5,000 and $20,000.[44]

In January 1981, just before Reagan's first inauguration, Lefever published an article of his own in the *Wall Street Journal*: "Politics and Baby

Formula in the Third World." He warned that the marketing code soon to be considered by the World Health Assembly had been drafted by "[a]nti-industry activists, UN bureaucrats and their allies" and that its adoption "would certainly be a victory for the activists who seized on the infant formula issue to establish a bold precedent for hobbling the multinationals through 'international regulation.'" At the upcoming Assembly meeting in May,

> [t]he atmosphere there, like that in the General Assembly, will doubtless be charged with Third World demands wrapped in moralistic slogans. Under these circumstances, Third World delegations and their Soviet bloc friends may well adopt the code with little examination. Western governments will be under pressure to go along to keep the faltering North-South dialogue alive. This will be an unprecedented attempt at international legislation by ideological intimidation.
>
> The die is not yet cast, and I venture to predict that the U.S. delegate under a new and more realistic administration will reject the compassion-coated revolutionary rhetoric and vote against the code and, at the same time, affirm our interest in infant nutrition in the Third World, the right of manufacturers to market their product and the responsibility of governments to adopt their own guidelines for foreign trade and investment.[45]

There is no reason to suppose that the thoughts Lefever articulated here were significantly at odds philosophically with those of the Nestlé leadership or with those of the corporate community as reflected in the editorial policy of *Fortune.* This is not to say that corporate interests and national interests—as seen by Lefever—have always coincided or that corporate leaders have never disagreed on questions of this sort or on political-ethical questions of other sorts. But these are the interpretive precautions always applicable to the case-study format.

We might look a little more at Lefever himself, taking pains not to identify him too closely with the corporate interests which he did evidently serve in at least this one instance.

Lefever was a realist in the fashion of Reinhold Niebuhr, some of whose works he edited,[46] and Hans Morgenthau, who introduced Lefever's 1957 book, *Ethics and United States Foreign Policy.* In that book, Lefever made several points relevant to our present discussion:

> Economic policies, like military policies, are designed to serve our overall foreign policy objectives. . . .
>
> The United States has an interest in raising the living standards and morale of the peoples in the less-developed areas of Asia and Africa, in order

> that they may become strong independent states and thus better withstand Communist attack or subversion. . . .
>
> Men who recognize the fundamental rights of people everywhere are deeply concerned with efforts to erase poverty, illiteracy, and disease. Freedom from want is a desirable goal in its own right, even if efforts in that direction do not make an immediate contribution to United States foreign policy objectives or to world peace. . . .
>
> The question then arises: Should the government of one nation have a humanitarian concern for the people under the jurisdiction of other national governments? . . .
>
> But a more difficult question arises: Should the foreign policy of a nation attempt to serve humanitarian objectives abroad? . . . The answer is both yes and no. The primary aim of foreign policy is to serve the national purpose and the interests which flow from this purpose. In pursuing these national interests, which are largely political in character, humanitarian concerns may also be served, but this is incidental.[47]

Lefever echoed the morality-versus-moralism arguments[48] made by Morgenthau and Kennan. He held that "[a] consistent and single-minded invocation of the 'human rights standard' in making United States foreign policy decisions would serve neither our interests nor the cause of freedom."[49] And he wrote on the ethics of covert intelligence methods, arguing at one point that American missionaries should not reject good opportunities to assist the Central Intelligence Agency, since they "are also U.S. citizens who have a stake in the survival of freedom. . . . [W]hen given a choice a conscientious policy-maker or citizen should choose the lesser of two evils or the greater of two goods."[50] Professionals, even religious professionals, should be willing to be state functionaries.

In May 1981, the Reagan administration decided to vote against the infant formula marketing code approved otherwise unanimously by those states voting in the World Health Assembly. The advice of President Reagan's close friend and future attorney general, Edwin Meese III, seems to have overridden contrary advice from other sources.[51]

This decision was extremely unpopular in the United States generally and elsewhere in the government. Both houses of Congress passed resolutions of protest and four different sets of hearings were called. Two prominent officials at the Agency for International Development, including the chief medical officer, Stephen Joseph, M.D., a pediatrician, resigned; the very same day,[52] hearings convened before the Senate Committee on Foreign Relations to consider the nomination of Ernest W. Lefever, Ph.D., to be Assistant Secretary of State for Human Rights and Humanitarian Affairs.

Lefever did not prosper before the Committee, despite its Republican majority. His shady dealings with and support for Nestlé counted heavily against him. His nomination was reported to the full Senate unfavorably, thirteen votes to four,[53] and he withdrew.

Though devious and disingenuous, Nestlé seems not to have violated international law or the laws of the nation-states in which it marketed infant formula. Its egoism was expressed mostly within the economic arena, where corporations "fight in a world of free enterprise."

If restrictions on the aggressive marketing of infant formula were so important and so widely supported, why were they not imposed directly by the nation-states whose babies were being abused? Why were such regulations needed at the international level? Why did the developing nations not reject international parentalism on this issue, as they have rejected it from time to time when Europeans or North Americans have tried to work their ecological wills on eroding hillsides and burning rain forests? These are questions whose answers lie beyond our subject.

XI

PRIVATIZING THE COMMON INHERITANCE OF HUMANKIND

AMBITIOUS national governments have characteristically tried to advance the prosperity of domestically based corporations active in international trade. The establishment, protection, or extension of trading rights and marketing opportunities has played a role in many violent conflicts—the Seven Years War, the Opium Wars, the Spanish-American War and Philippine Insurrection, World War I, the Pacific War—as well as in countless nonviolent conflicts. As we have just seen, resistance to the international regulation of corporate activity underlay American opposition to restrictions on the marketing of artificial infant feedings in the developing world; it mattered little that American firms were only peripherally involved in the inciting dispute.

A parochial interest in the success of trading companies, whether domestically based private corporations or national trading companies, will surely outlast many other features of international politics. And this interest will most likely continue to be expressed in the realist or "nationalist" or "mercantilist" metaphor.[1]

In the past, unless they were directly involved in the international trade of items like pharmaceuticals or advanced seed grains or unless they were trying to change specific policies (like the promotion of tobacco exports from the United States to countries with weak public-health education programs),[2] life scientists have had little to do with these matters. And, for the most part, relevant government policies have been comparatively restrained; medicine and food have less often been tools of national advantage than have other products.

In the future, though, many more life scientists will be involved in international trade issues, sometimes reluctantly, sometimes willingly, sometimes willfully. The proximal reason is extension of intellectual property-rights principles into areas of *basic* scientific knowledge. The deeper reason is redundancy of "basic" and "applied" questions in modern life-sciences research, such as in genetic engineering, where fundamental mysteries and practical problems may have identical solutions.

Life scientists have had trouble with this new ambiguity, and so have jurists. United States case law has extended patent protection to the "inventors" of genetically novel organisms, such as "engineered" bacteria and animals, the latter potentially including primates. The landmark precedent in this regard has been *Diamond, Commissioner of Patents and Trademarks v. Chakrabarty*, 1980.

Chakrabarty and its sequelae have transformed international trade prospects in the restriction-enzyme era. In turn, they have reinvigorated a classic debate in political ethics.

In the early 1970s, Dr. Ananda M. Chakrabarty, an employee of the General Electric Company, introduced into bacteria of the genus *Pseudomonas* sets of foreign genes imparting a potentially marketable ability. Chakrabarty's organisms could metabolize simultaneously "at least two"[3] of the many hydrocarbon components of crude oil.

Chakrabarty had "bred" these new bacterial strains, not "invented" them in any traditional sense. He had not even genetically "engineered" them in the now-usual sense; genetic engineering, or "gene splicing," had not yet been described at the time he was working.

What Chakrabarty had done was interesting but not revolutionary. He had isolated bacteria independently able to metabolize at least one crude oil component. He had then incubated these different types of bacteria in the same culture medium, in which medium eventually there evolved several strains able to metabolize more than one crude oil component at the same time. Some of these more competent bacteria were then subcultured and characterized. The desired evolutionary changes had been made more likely by various laboratory manipulations, but directed genetic recombination had not been attempted or achieved.

The crude-oil-metabolizing properties of interest depended on enzymes whose structure was encoded by genes contained in "plasmids." Like viruses, plasmids themselves are not living things, not full organisms. Plasmids are circular molecules of deoxyribonucleic acid, the genetic material, DNA. They are extrachromosomal. In other words, they exist outside a host cell's collection of species-specific genes, its characteristic set of chromosomes, its "genome." Many different sorts of plasmids exist, and plasmids of various types are found in most bacterial species.

If one group of bacteria harboring a particular plasmid is properly mixed and incubated with a second group of bacteria not harboring that plasmid, plasmid transfer may occur.

For instance, some plasmids code for enzymes that inactivate antibiotics; as a class, these plasmids are called "R" plasmids—"R" for "resistance" to antibiotics. Consider bacteria from three unrelated colonies: R-plasmid-containing bacteria resistant to penicillin, R-plasmid-containing bacteria resistant to tetracycline, and bacteria which contain no R-plasmids and which are resistant neither to penicillin nor to tetracycline. Now consider incubating a mixture of bacteria from each of these three different colonies. One might thereby produce three different types of bacteria *each* resistant to *both* antibiotics.

An analogous phenomenon occurs randomly in nature—whence, in large part, the clinically troublesome co-evolution of different bacterial pathogens all resistant to the same antibiotics.[4]

Plasmids would soon become favorite vehicles of genetic engineers. They would be spliced and recombined in and into sequences not found in nature. But the plasmids Chakrabarty used had not been so altered. Chakrabarty's achievement depended ultimately on the random migration of useful plasmids into individual bacteria. His organisms—two strains were being considered together—had been "invented" only in the sense that they had been incubated with plasmids of interest and had been induced to become their hosts, thereby gaining the advertised ability to "eat" as many as four components of crude oil. Inconveniently, though, crude oil contains many more than four components. And, as we will see, Chakrabarty's organisms proved less "stable" than his application claimed.

At any rate, in 1972, General Electric requested patent protection, first, for the process by which Chakrabarty's bacteria had been produced, second, for an oil-spill clean-up technique into which they were to be incorporated, and, third, *for the organisms themselves.* Louis Pasteur had received a United States patent in 1873 on a beer fermentation process requiring a "manufactured" yeast;[5] subsequent patents had been granted on vaccines containing "active" biological agents, such as intact attenuated viruses. General Electric's first and second requests seemed within these bounds and were granted. Its third request was denied. Microorganisms were "products of nature," said the Patent Office, not inventions, and living organisms such as those presented were not patentable according to statute.

General Electric requested a review by the Patent and Trademark Office Board of Appeals. Chakrabarty's organisms were held by the Board *not* to be "products of nature," since their evolution had been arranged and since identical, "discovered" organisms had not previously been described. Despite this notable concession, the Board, bothered by the vitality of these "products of man," confirmed the nonpatentability of the organisms according to statute.[6]

At this level, the application turned on two laws. One was the Plant Patent Act of 1930, which had been designed to encourage the industrialization of asexual plant breeding, the type of plant breeding popularized by Luther Burbank, whose improved potato had become widely accepted in the nineteenth century. Also figuring was the Plant Variety Protection Act of 1970, which extended protection to sexual plant breeding through a system of "certification." This latter Act specifically excluded from certification "fungi, bacteria, or first generation hybrids" and exempted researchers and farmers from commercial obligations to breeders. The specificity of the two Acts suggested to the Patent Office and to the Board that Congress had not and did not consider living things patentable as a general rule and that Congress definitely considered bacteria unpatentable, at least as of 1970.

When General Electric appealed the Board's ruling to the United States Court of Customs and Patent Appeals, two of five judges agreed with the patent refusal, but three did not agree, finding that the Acts should have been considered irrelevant to the application. Patents could now be obtained on invented living things themselves, not just on processes of production or use.[7]

When the same five judges were asked to reconsider a year later, the reversal stood, this time four-to-one, the judge writing for the majority denying that the decision would "extend" the patent laws: "'The sky is falling, the sky is falling!' cried Chicken Little. . . . Come, let us return to reason."[8]

This time it was the Patent Office that appealed, but unsuccessfully. Five of nine Supreme Court Justices found the lower court's arguments persuasive, agreeing that routine criteria accommodated Chakrabarty's organisms and agreeing "that the fact that micro-organisms are alive is without legal significance for purposes of the patent law."[9] In affirming the Appeals Court, the Supreme Court concurred that patentability did not require the Acts of 1930 or 1970, nor was it precluded by any of their provisions.

Apparently, the Supreme Court's disinclination to apply the Acts to *Chakrabarty*, one way or another, had nothing whatever to do with the vast technological difference between, say, the grafting of rose bushes and the introduction of foreign genetic material into a host cell. Just as apparently, there is nothing in the decision indicating that the Court seriously anticipated consequences ethically unlike the ones narrowly at law. Certainly, Chief Justice Warren Burger, writing for the majority, did not consider such potential consequences relevant to the issue. Indeed, he was rather dismissive of long-term consequences generally:

> We are told that genetic research and related technological developments may spread pollution and disease, that it may result in a loss of genetic diversity, and that its practice may tend to depreciate the value of human life. . . .
>
> It is argued that this Court should weigh these potential hazards in considering whether respondent's invention is patentable subject matter under [patent law]. We disagree. . . .
>
> What is more important is that we are without competence to entertain these arguments—either to brush them aside as fantasies generated by fear of the unknown, or to act upon them. . . .

> We have emphasized in the recent past that "[o]ur individual appraisal of the wisdom or unwisdom of a particular [legislative] course . . . is to be put aside in the process of interpreting a statute." *TVA v. Hill, 437 U.S., at 194.* Our task, rather, is the narrow one of determining what Congress meant by the words it used in the statute; once that is done our powers are exhausted.[10]

The Plant Variety Protection Act of 1970 "excluded bacteria from its protection" specifically. Nonetheless, held the majority, the Act "does not evidence congressional understanding" that living things such as bacteria were unprotectable.[11] We may assume that the chief justice considered the exclusion somehow immaterial, but not unconstitutional, not "bad law." At least, we have no record of a request to his associate justices that they join him in overturning the Act in part or in whole. If the Act was valid, then the manifest inconsistency of statute and holding suggested that Chief Justice Burger had chosen *not* to "put aside" his "individual appraisal of the wisdom or unwisdom of" this "particular [legislative] course."

Justice William Brennan, writing for the minority, did not join the broader issue but argued that the Acts were clearly applicable, that the Patent Office had correctly followed congressional intent in denying the application, and that the patenting of "invented" living organisms should not be allowed without new and specific congressional action.

Once out of its jurisprudential Petri dish, *Chakrabarty*, as case law and as legislative and regulatory mandate, proved elastic and provocative. The Patent and Trademark Amendments of 1980 and 1984, the Technology Transfer Act of 1986, and certain executive orders signed in 1983 and 1987[12] all accelerated the patenting of what many had assumed to be public domain. Or, in a less parochially American sense, these measures all accelerated the "privatization" of what was arguably the common inheritance of humankind.

In 1987, the Board of Patent Appeals approved the patenting of unique, "man-made" animals, starting with an oyster augmented with ten extra chromosomes.[13]

Walter Gilbert, Ph.D., simultaneously of Harvard University and the Genome Corporation, spoke ambitiously of "copyrighting" human genomic sequences, making them "available to everyone—for a price."[14]

Harvard's in-house patent attorneys postulated the patentability of the glycoprotein 120 antigen (GP120), the most immunogenic protein on the "coat" of the human immunodeficiency virus (HIV), the cause of AIDS. They succeeded in getting it patented, even though it was a naturally occurring protein discovered on a much-studied protein coat.[15]

In *Chakrabarty*, Chief Justice Burger had written that "a new mineral discovered in the earth or a new plant found in the wild is not patentable subject matter."[16] What would he have recommended in the case of GP120? It was not an "art," not a "process," not a "machine," not a "manufacture," not a "composition of matter," not "man-made," not truly "nonobvious" within the international virological community, and definitely not "new."

But it was "useful." Harvard awarded an exclusive GP120 license to Cambridge Bioscience Corporation, which had previously given Harvard a $350,000 grant. Harvard said it would use its GP120 royalties to fund fellowships for AIDS researchers in developing countries.[17]

On April 11, 1988, the United States Patent and Trademark Office announced final approval of Harvard's application for proprietary rights to a transgenic mouse.[18] The approved patent claimed "a transgenic nonhuman eukaryotic animal (preferably a rodent such as a mouse) whose germ cells and somatic cells contain an activated oncogene sequence." The rights granted went well beyond the celebrated mouse and its clones or progeny, taking in any "transgenic nonhuman mammal all of whose germ cells and somatic cells contain a recombinant activated oncogene sequence introduced into said mammal, or an ancestor of said mammal, at an embryonic stage."[19]

Considering the long-acknowledged feasibility, even the inevitability, of the achievement, a rather large slice of life was being placed on a single, privately held plate. Scores of public and tax-exempt institutions and generations of publicly funded scientists from the "pre-patent" era did not share in these propriety rights to transgenic nonhuman mammals, rights which Harvard quickly licensed by prearrangement to E. I. DuPont de Nemours and Company,[20] a major sponsor of the "Oncomouse" project.[21]

A Patent Office spokesperson, apparently hoping to preempt public concern, volunteered the government's intention not to consider applications protecting novel "human" organisms. However, public concern was not preempted, nor were attempts to legislate a moratorium on the patenting of life, especially animal life.

In the matter of Chakrabarty's application, the Patent Office, the lower court, and the Supreme Court had debated the extension of patentability from plants of agricultural and horticultural interest to novel bacteria—by popular taxonomy, all within the same "kingdom," the plant "kingdom." Plant-breeding statutes had been felt enabling by some, limiting by others, and irrelevant—or inconvenient—on final appeal. But the patentability of "plants" had been the issue, not the patentability of animals. Technological details had been regarded as material to the question of originality but incidental to the question of long-term ethical consequence.

Now, with the patenting of transgenic nonhuman mammals, *Chakrabarty* had extended the potential reach of intellectual property rights in the life sciences not just from plants to other "plants," as had been thought reasonable by the Court, but from plants to animals as well. Not because bacteria and oysters or mice were in the same "kingdom" but, apparently, because they had become subject to similar technology. Similar enough by *lay* standards, we should say, though not at all similar by *laboratory* standards. Sensibly or not, the technology employed, not the taxonomic boundary crossed, seems to have become the new informal test governing patentability extension.

The distinction between invention and discovery, the distinction between "things" and "beings," was fading fast. By the late 1980s in the United States, the number of scientists, corporations, industry-university consortia, and federal agencies applying for patent protection had far exceeded the government's ability to respond.[22]

Purely technological advances gained protection. But so did many discoveries of a type formerly considered to be "basic science" and therefore unpatentable, discoveries of a type formerly shared freely within an international community of colleagues—almost all of whom, in the United States as elsewhere, had been educated and trained substantially at public expense, most of whom were still salaried by public or publicly supported or publicly supplemented or publicly favored institutions.

Incredibly, at one point in 1989, the NIH was refused a sample of a putative "virus-like infectious agent"[23] newly isolated from seven AIDS patients by investigators at the Armed Forces Institute of Pathology (AFIP). The discovery, it turned out, was the subject of a patent application, and only certain "collaborators" were being given full access to details. With whom was the AFIP collaborating, if not the NIH? "We're keeping that under wraps," said the AFIP's director, a Navy captain.[24]

Ironically, Chakrabarty's precedent was to outlast his invention. His protected organisms, at best, had metabolized just a few of crude oil's many hydrocarbons. Their several progenies were even less effective. The reason for this generational decline was fundamental. His organisms, contrary to claim, were not genetically "stable."[25] Daughters separating from patented parent organisms too often lacked promised genes or functioned in new and undirected and unpatented ways. And, beyond the question of efficacy, genetically unpredictable organisms could not be considered safe for environmental use. This was disappointing for General Electric. And it must have been uniquely embarrassing for the Court (if any of its ever-changing clerks were still paying attention) and for the Patent Office. As if transistors were "dis-inventing" themselves into vacuum tubes.

For a mix of reasons, policy did adjust. No new animal was patented from April 11, 1988, till December 29, 1992, when three more transgenic mice were certified as inventions, but more cautiously and less comprehensively than before. During this undeclared "moratorium" of four-and-a-half years, this "natural experiment" in science-and-technology policy, research did *not* end—"more than" 180 applications accumulated—and commercialization did *not* cease. Similar research and development proceeded apace in Europe, despite the fact that the European Patent Office had never issued an animal patent, and, if it heeded a December 1992 recommendation of the European Parliament, where environmentalists and animal-rights advocates had become entrenched, it never would.[26]

Should Chakrabarty's patent-on-life have been granted in the first place? Should patents have been granted subsequently on other living things or on viruses or plasmids or on "found" pieces of nature, altered or not?

Yes, according to the United States Constitution and United States statutes as interpreted by an Appeals Court, by the Supreme Court, and, following judicial direction, by the Patent Office. And, yes, according to those anxious to establish an American lead in biotechnology by assigning property rights to risk-takers willing to enter a new field creatively.

But, no, according to many others. Objections have fallen into diverse categories, one of which contains arguments against "privatizing the common inheritance of humankind." We will now discuss a subcategory of "privatization" particularly vexing to many in the international community.

Nationalizing Biotechnology

Do citizens of scientifically backward nations have a morally sensible extra-commercial claim to the products of life-sciences research and development? Should new understanding in the life sciences be considered the "common property" of humankind? Is there an international "right to charity" in biotechnology? Less aggressively, is there an international "right to technology transfer" in the life sciences?

These questions have a longer history than might at first be thought. While judging their competing answers, we must keep in mind a story of interregional exploitation, transgenerational resentment, and insincere cooperation.

In countries decolonized after the end of World War II, prosperity followed independence neither often nor quickly. Many activists and some scholars had thought that it would, especially where natural endowments were rich. But, even in fortunate countries, the lessons of economic and political development proved harsh. The export of low-priced commodities and the provision of unskilled labor could not predictably advance agricultural living standards, and masses of poor farmers meant crowds of unemployed city dwellers, all needing to eat food they could not afford.[27] Solutions were proposed, and each was tried somewhere at sometime. By the early 1970s, none seemed reliable.

Then, starting in 1973, the Organization of Petroleum Exporting Countries (OPEC) made a staggering success of one special development mechanism: the commodity-pricing cartel. OPEC's methods were not generalizable, however; copper and tin and coffee and bananas and sugar and rubber were not oil, as the developing world would learn to its dismay. Still, for a time, commodity-pricing cartels seemed a trick worth pulling.

Hope now acquired a wishful name: the New International Economic Order, the "NIEO," envisioned by seventy-seven developing nations—the original "Group of 77"—and proposed as a goal at the Sixth Special Session of the United Nations General Assembly in 1974. Major objectives of the NIEO dealt with commodity pricing, economic sovereignty, debt relief, and the like; lesser objectives were numerous, some reasonable, some peevish.[28]

The developed world, stung by OPEC, was not in a concessionary mood, and the Group of 77 proved punchless. It was never seriously heeded, and its demands for the NIEO became less frequent and less fervent. Commodity cartels, even for oil, proved hard to manage, and commodity prices plunged during a worldwide recession. Debt became the all-consuming problem for chronically poor and suddenly poor nations alike. Finally, a star surpassing in brilliance even the NIEO arose in the East: export-led growth, a strategy made famous in the newly industrialized economies of South Korea, Taiwan, Hong Kong, and Singapore.

So much for the NIEO, save three legacies affecting our subject.

The first legacy derives from a claim that "the common inheritance of humankind" comprised specific, listable assets. The best-known claim listed the sea bed's mineral resources, such as manganese nodules, to which even land-locked nations asserted co-ownership, lest sea-faring states and their giant corporate agents monopolize them. The much-troubled 1982 United Nations Convention on the Law of the Sea was the outstanding result of the "common inheritance" argument.[29]

This first idea recalls natural-law traditions and importantly overlaps Lockean liberalism and, congruently, life-sciences liberalism. But including the sea bed within "the common inheritance of humankind," while logically feasible, was neither immediately convincing nor ultimately self-justifying. Even Locke, noting its size and the amount of labor needed to mine it, would likely have thought the sea bed an ideal natural commons from which to extract private property. By vital contrast, maintaining the genetic code and the rules and means for its manipulation within a category called "the common inheritance of humankind" is instantly compelling on the three firm grounds of commonality, heritability, and humanity.

The second legacy derives from an unapologetic insistence that international wealth be redistributed. Rich states had mercantilized and industrialized at the expense of the colonized world, and they were acting similarly in the postcolonial era through transnational corporations. By this analysis, OPEC's outrageous pricing policy was a model of fair retribution, not extortion. Regrettably, OPEC's pricing policy applied even more cruelly to poor states than to rich ones. But more sensible, nonretributive forms of redistribution were also put forward: debt forgiveness was one, technology transfer another.

This second idea, like the first, resonates with life-sciences liberalism, at least on a theoretical note. Locke's arguments for surplus redistribution and for charity were—and remain—emotionally consonant with most every contributor to the tradition, if not since Scribonius, at least since Bacon.

The third legacy derives from the assumption that technological change could serve—and the assertion that it *should* serve—as a prime stimulus to economic growth in developing countries. The importance of technological change had long been recognized in *developed* economies, though the true size of its effect, exceeding that even of capital, was grudgingly accepted at first; Robert Solow's agitating work was thirty years old in 1987 when blessed by the Nobel Prize.[30] The success of export-led-growth economies made the design and manufacture of internationally competitive goods an all-too-automatic goal. But the developing world could not afford for ever to rely on the inconstant attentions of visiting scientists and technicians serving the needs of advanced foreign economies. A permanent solution to the impoverishment problem required local acquisition of locally appropriate technological competence. As suggested by agriculture's "Green Revolution," "exogenous" technological inputs had to be made to some significant degree "endogenous."

This third idea inhabits the scientific tradition generally, and a variant resides in the life-sciences tradition specifically. Technological change propels the advancement of human welfare, if not economic growth, and technological changes have their best and safest effects when understood by those who employ them. Thus, for instance, have medical missions and aid programs sought sequentially to educate medics, technicians, nurses, midwives, physicians, and research scientists. Thus, as well, did the United Nations Biotechnology Center come into existence, however fragile financially and administratively and however unpromising politically.[31]

These three ideas have deeply affected international debate on matters of distributive justice. New and disruptive in political practice, yet old and restorative in the political ethics of the life sciences, they apply with special force to problems in international biotechnology policy.

Within the normal flow of international trade, international teaching, and international assistance, innovations in the life sciences would disseminate predictably to most areas, though slowly to some. Pharmaceutical houses and equipment manufacturers and seed companies would charge what they could for their goods, more during periods of market advantage, such as patent protection, less so at other times. Health ministries and hospitals and farmers' cooperatives would have to compromise on matters of quality and convenience and efficiency. International teaching and assistance would help significantly, but many people still would suffer and die only because they were born into poor or primitive societies. Although unsatisfactory on an absolute scale, biotechnology dissemination would meet within this normal flow no barriers *intended* to stop it or even to slow it down. Such barriers would seem to serve neither commerce nor compassion.

Who would erect such barriers, why would they be raised, what would they look like, and just how could they be assembled?

Barriers could, of course, be erected corruptly, as when a Japanese pharmaceutical firm apparently bribed a scientist-administrator to retard the evaluation and marketing of heat-treated HIV-free coagulation factor from rivals in Europe and the United States[32]—a "North-North" problem. Minus corruption such as this and minus primary corporate collusion, dissemination barriers would most likely be erected intentionally by states. They would be raised presumably to protect or advance national trading interests, though they might be excused on other grounds. They would probably be "informal" barriers, nearly transparent to the General Agreement on Tariffs and Trade (GATT): refusing to consider foreign clinical trials in new-drug approval procedures, expediting back-logged patent applications according to an unwritten "home-county-first" rule, delaying the introduction of an advanced foreign product soon to be challenged by a domestic competitor—as when *unbribed* French blood-banking officials in 1985 knowingly disseminated HIV-laden coagulation factor rather than allow a safe American product to preempt a hoped-for domestic product.[33] In the United States, these barriers could best be assembled from two familiar civic stockpiles: "industrial policy" and "international fair-trade policy," better known jointly as "competitiveness."

In December 1987, in the United States Senate, Lawton Chiles of Florida introduced "The Biotechnology Competitiveness Act." In June of 1988, the Act passed the Senate, eighty-eight to one, William Proxmire of Wisconsin the lone dissenter. In the House of Representatives, Chiles's bill evolved into "The Biotechnology Science Coordination and Competitiveness Act" and proceeded to joint hearings before two subcommittees of the House Committee on Science, Space, and Technology on July 14, 1988. Realist thinking predominated.

Testimony and comment centered on the risk to "economic national security"[34] posed by the rise of life-sciences powers able to challenge or frankly to end American dominance in a field long commanded by the United States. Farsighted Japanese industrial policymakers had "targeted" biotechnology in 1981 and their investments were now being realized.[35] American projections placed yearly international trade in biotechnology at $40 billion by the year 2000; the Japanese projection was $100 billion.[36] More startlingly (and less credibly), the Japanese were predicting that biotechnology would represent 11 percent of their gross national product by that same year.[37]

Biotechnology was still generating a trade surplus for the United States while wide deficits in other critical areas were growing ever less tolerable.[38] If we Americans did not act quickly to set "our own national strategy for biotechnology,"[39] then this industry too would follow the path of consumer electronics and the computer, with terrible consequences. The United States would be yielding to Japan and to a united Europe "the core technology for the next industrial revolution."[40]

Maddeningly, while all Americans were engaged "in this global race against time to assure our eminence in biotechnology," the Patent Office was years behind.[41] In excess of 7,000 biotechnology patents were pending. Separate hearings[42] had been held on this single issue four months before, but improvements were not in evidence. The Japanese, some suggested, were compounding the problem by filing trivial or unfair claims in great numbers, hoping to swamp and confuse the Patent Office, thus slowing work on legitimate applications filed by Americans.[43] Worse, they and most other competitors were "pirating" the "process" patents that were supposed to be protecting American innovations in bioindustrial engineering.[44]

Advancing American biotechnology was a nonpartisan "national survival" issue.[45] More government involvement and more free enterprise were needed simultaneously.[46] Everything patentable should be patented as quickly as possible, including the maximum number of private claims resulting from government-funded research.[47] As "the first order of business," urged one witness from industry, genome mapping should be privatized[48] and human genome sequences themselves should be considered patentable.[49]

In these proceedings, the special political-ethical status of biotechnology escaped attention nearly altogether. There was one, brief, ethically sensitive prepared statement,[50] but it seemed distinctly out of place. Hour after hour, page after page, biotechnology was a sword-in-the-stone to which the United States had a transcendent right of retrieval and with which it must subdue its ignoble rivals, thereby protecting the succession of its progeny to a just inheritance. Old epic, new magic.

How telling was the term "biotechnology competitiveness." Biotechnology immunized, healed, and fed. But biotechnology, we were now to accept, could also enrich and impoverish, reward and forsake. It could still secure individuals from paralysis, pain, and want. But it could also be made to secure one nation from its rivals and rich nations from the poor.

Less dismally, was it possible that privatization in its two major forms, commercialization and nationalization, was in truth a glove well-fit to that great invisible hand said to guide every free economy, local or global? Was it possible that all would benefit from the greed of each? Yes, however unlikely, it was possible.

That said—and it has never been said *convincingly* in the case of biotechnology—a legislated moratorium on "basic" patents might have been a better idea than accelerated privatization, even as late as 1988.

For-profit biotechnology firms, we should remember, had existed well before *Chakrabarty* came to final judgment. Cetus had been started in 1971; Bioresponse in 1972; Genentech in 1976; Genex in 1977; Biogen, Hybritech, and Collaborative Research in 1978; Molecular Genetics and Monoclonal Antibodies in 1979; Calgene in 1980.[51] True, they had generally favored the extension of patentability; Genentech had even filed as an *amicus curiae* in *Chakrabarty*, urging the Supreme Court to affirm the Appeals Court ruling.[52] These new biotechnology firms had initially been capitalized in the intellectual property-rights environment shared by pharmaceutical, biochemical-engineering, agricultural, food-processing, and brewing companies. If *Chakrabarty* had gone the other way, they and their competitors would have adapted. Or, more accurately, they and their competitors *would not have had to adapt. Chakrabarty* ended an era as much as it began one.

By 1988, many biotechnology firms were competing not at all on the basis of price, service, quality, or applications innovation. They were competing only for a salable piece of basic science, competing only for monopoly rents sure to distort the future course of research and development—academic, industrial, domestic, and foreign. Hardly the conditions of "perfect competition" and hardly the most efficient way to manage a technological chain reaction. Or to foster North-to-South transfer of appropriate technology.

More so in the life sciences than in any other field, public and publicly favored institutions had been marvelously productive. Though they had sometimes been underfunded, there was no believable evidence that they had been slothful, that they needed replacement by private entrepreneurs. If *Chakrabarty* had gone the other way eight years before, and if major additional public funds had flowed to universities and institutes eager to expand their pursuit of *new* research goals in response to *old* but, evidently, adequate incentives, progress might actually have been quicker—surely less litigious[53]—and its end effects fairer.

But in these 1988 hearings, the better-tested alternative—public funds for publicly responsive, public-domain science—was not the "competitive" favorite. It did, however, make one sudden, nostalgic, and instructively awkward appearance.

After a great deal of semi-chauvinistic testimony and comment, several professional witnesses were discussing the "wonderful opportunity" for "international scientific cooperation" presented by the mammoth effort just then underway to map and sequence the entire human genome. "In fact," it was said, "it just doesn't make any sense to try to have a gene's location known by one country and not by another. It's against the norms of scientific discourse." And so forth.[54]

Then, after almost a full day of refusing to choke on Japan's "exhaust gases . . . in the field of biotechnology,"[55] the chairman of the joint subcommittee, Congressman James H. Scheuer of New York, had a liberal epiphany:

> Mr. SCHEUER. . . .
>
> This whole question of what we hoard and what we share is enormously complicated and a terribly important question that we're not going to settle in the course of this hearing, and that will be one of the major roles of the commission [the proposed Biotechnology Science Coordinating Committee], to sort of parse out. But can any of you describe some of the places where we might draw the line? For example, can we do joint projects with the Europeans and with the Japanese on basic research and then say, "All right, when it comes to applications, we're all going to share in this pool, and that's where the competition is going to be"?
>
> I say this because this isn't a technology for just producing a better car or a better mousetrap or a better squawk-box, a better consumer electronic product. A lot of these applications of biotechnology in the field of health and the improvement of agricultural products of all kinds will produce enormous benefits in the third world; crops, for example, that will withstand drought and insects, that will require less irrigation, less fertilization.

I mean, there's a real global public interest in getting this stuff out there and in use. It's not just purely a private enterprise business. It matters little whether the latest VCR comes sooner or later. Over the fullness of time, mankind will not measure that as very important. But whether we develop some of these new plant and animal species and some of these new health applications is of considerable importance to mankind and will save, perhaps, millions of lives and enhance the quality of life, especially for people in the developing world if we can get this basic information—this basic, pure research—out there, and then let the competition be on the applications.

Does this make sense? Is there any other formula that you would devise that would provide a rational program of where there should be data-hoarding—first, let's say where there should be data-sharing and joint involvement, and at what point we say, okay, we've done the basic research, now get on with it and let's have the competition on the applications? Is there a better way of parsing it out?

There ought to be a way of saving time, and therefore saving lives, and improving the quality of life in all these areas for people around the world. The Japanese have a lot to contribute to us. The Europeans have a lot to contribute to us. We probably are still first, although that lead may be faltering somewhat. We have an enormous amount to contribute. Isn't this a case where, at least for basic research, the whole is greater than the sum of the parts? We ought to find mechanisms for collaborating and then say, all right, this is for applications, machines, tools, get on with it and do your thing?[56]

The pace of life-sciences progress had made less clear the distinction between basic and applied research. Yet, that was merely a complication, not the real issue. Biotechnology could never be the political-ethical equivalent of other technologies and was ethically unsuited to international competition. In the life sciences, globally minded cooperation among advanced states was a better idea than competition, but it was a challenging idea requiring cautious thought.

Congressman Scheuer, to his credit, was having a vision of the true problem. But his liberal reverie was interrupted by a specious argument from analogy. Hopefully, an *innocently* specious argument from analogy:

Mr. [*sic*] NELSON [Dr. David B. Nelson, Executive Director, Office of Energy Research, U.S. Department of Energy]. I certainly agree with you. I would add, though, that this is complex. I realize—

Mr. SCHEUER. Of course it's complex.

Dr. NELSON. But let me give you a specific example of the complexity.

Dr. MERRIFIELD [Dr. D. Bruce Merrifield, Assistant Secretary of Commerce for Productivity, Technology, and Innovation] testified earlier about the Superconductivity Competitiveness Act that the [Reagan] Administration has introduced. One of the provisions of that act is a modification of the Freedom of Information Act, FOIA. You might ask, what do those have to do with each other? But the answer is that if we do joint work—for example, in our National Laboratories—between Government funding and private funding, the data that results from that joint work—and in the case of industry funding, it may be leading to what they would want to have as proprietary data, because it leads to a product—that [*sic*] data may be releasable. We don't know, but we think that it may be releasable under an FOIA request.

Now, FOIA is an important part of our laws. It has safeguarded the openness of our political or governmental system. But here is an application of FOIA that may frustrate efforts for private sector and Government funding to work together toward near-term commercialization, where proprietary data could be an important product.

This is just one example of the complexity of determining where the proper boundary is between openness—which is for all, but for no one in particular—and closedness which leads to commercialization, because I have something that I can sell; I own it and you don't.

Mr. SCHEUER. I totally respect that dilemma that you've given us. That is a conundrum of massive proportions.

Dr. NELSON. You'll have to deal with it because I believe this bill has been introduced, so Congress will have to take up this question of modifying FOIA.

Mr. SCHEUER. Of course, we have to deal with it.[57]

Dr. Nelson's agency, the Department of Energy—maker and tester of nuclear weapons, keeper and leaker of radioactive waste, proprietor of the National Laboratories, such as Los Alamos and Lawrence Berkeley, and expert on the genetic effects of radiation—had lobbied more insistently than any other organization, public or private, for comprehensive mapping and sequencing of the human genome, to the early confusion and concern of many observers, and had taken independent action to begin the project. The commercialization of superconductivity was an enterprise in which the Department's interest had required less explanation, but it was ethically quite unlike the genome project, quite unlike the granting of proprietary rights to human nucleic acid sequences.

Was the Department of Energy, as feared, viewing life-sciences policy through its built-in big-physics, big-industry, national-security lens?

Maybe not, but having to win public access to life-sciences research data through the Freedom of Information Act was an image, intended or accidental, that deserved permanent storage in a containment building for poisonous ideas.

Anyway, Congressman Scheuer was not quite finished. Having seen the true problem, he now proceeded to ask a premature and ill-formed version of the right question:

> What I'm suggesting is that there ought to be a way of investing real resources, and maybe this is the part where the Government does funding for the fundamental research, and then turns the private sector loose. When I say "Government," I mean all these governments.
>
> There is talk about burden-sharing in terms of our military defense. The Japanese are spending no more than one percent of their GNP; it's a constitutional limitation that we provided them gratis, coming from our own Government right after the end of the war. We spend 6.5 percent, so there has been talk about burden-sharing.
>
> Isn't this a way that there could be some burden-sharing? Perhaps we could suggest to the Japanese that there would be an asymmetrical contribution to a joint fund for basic research, part of which they would consider their contribution to burden-sharing. After all, a lot of these new developments are going to be of enormous benefit to the third world in a whole variety of ways. This could be their form of burden-sharing; and we would contribute something, perhaps asymmetrically, and the Europeans would, too. Then when it got to the point of applications, then the collaboration would end and private competition would commence.
>
> Is that a conceivable formula?
>
> [No response.]
>
> Mr. SCHEUER. Who knows? We have to discuss it.[58]

Could basic-research contributions to the American life-sciences community and biotechnological contributions to the welfare of the developing world be accepted as substitutes for military contributions to Western force posture?

"Is that a conceivable formula?"

No.

A similar scheme may earlier have metamorphosed into Japanese Prime Minister Nakasone's "Human Frontier Science Program," an unexampled piece of life-sciences internationalism proposed to faint enthusiasm at the 1987 Economic Summit in Venice.[59] Its brewers-become-biotechnologists notwithstanding, Japan's most reliable failing was life-sciences research; its first and only Nobel laureate in medicine-or-physiology left the country. Japan's best strategy, without question, was international cooperation, even if it had to be yen-powered.

But cooperation was Japan's problem. America's best strategy was competition.

A change of subject was in order, and Congressman Scheuer was soon returned safely to the ominous mess in the Patent Office: "'for the want of a nail, the empire was lost,'" as he put it.[60]

The House version of this bill never came to a vote. The Reagan administration opposed it on grounds that its provisions would either duplicate or complicate *ad hoc* efforts already far advanced.[61] The nationalization of biotechnology, a newly surveyed sector of the common inheritance of humankind, was underway.[62]

XII

ADVANCING THE PUBLIC HEALTH

THE MOST CURIOUS FEATURE of the market for human life is volatility in the terms of trade. We may imagine that regulation of this market was a founding goal of all cultures and civilizations, for it is now among their signal, if subtler, functions.

In politically developed societies in a nominally slaveless era, human life is popularly assumed to be unpriced. It is also assumed—or said—to be priceless. The first assumption seems correct only because the market price of human life is volatile: nearly zero for the next victim of heavy artillery or out-of-sight starvation or adolescent gang violence but whatever-it-takes for one child in a well shaft[1] or, perhaps on the basis of honorary humanity, three whales in an ice flow.[2] The second assumption, that human life is priceless, is manifestly false.

Human life and its desirable attributes—health, happiness, self-determination, and so forth—are goods in a barter trade whose more regular patterns long ago became institutions: violent intergroup competition, in which life is traded for tactical advantage or certain lives are traded for certain others; retribution for criminal behavior, in which life or liberty cancels a social debt; quarantine or the exercise of other extraordinary public-health powers, in which private rights (sometimes including the right to seek medical care) are suspended for the greater good.

In the formation of health policy, the rendering of this barter trade into a stable and more-or-less equitable form of exchange has become a focus of nearly reverential effort and, in the United States in the late twentieth century, sharp controversy, for it is here that the ethic of public health and the technique of "domestic" utilitarianism have conjoined, almost as twins.

In the ethic of public health, the individuals and groups from whom sacrifice is requested or compelled must be identifiable victims of bad luck or their own foolishness; their sacrifice must be minimized, though it cannot always be minimal; it must be honored as intentional and regretted as exceptional, even when neither. The ethic of public health is faintly military, and actions judged proper by its standards may be realized through police power.[3]

"Domestic" utilitarianism—the common variant of utilitarianism operating only *within* societies, not between or among them—is similar but not quite the same. Those who sacrifice need not be consulted or informed or even identified, though they might know who they are or who they will be; they need not be victims, though they might, arguably, be *made* victims; their sacrifice must be minimized, though it cannot always be minimal; it usually carries no honor, though, officially, it must always and sincerely be regretted. "Domestic" utilitarianism is efficiency made manifest, and actions seen to follow from its assumptions may be supported by all good citizens and may be realized through civil and, if need be, criminal codes.

The combination of these two ethics has been smooth and, for some, gratifying. The chief result has been the belief that ethical tension inextractable from health-policy dilemmas can be relaxed *in situ*. The mechanism of this détente has been (and must apparently be) classical utilitarian moral calculus and neoclassical economics and allied methods: welfare economics, cost-benefit analysis, industrial-organization theory, operations theory, path analysis, game theory, formal decision analysis. The consequence for health-policy reform in the United States has been the well-meaning and usually unconscious redescription of the patient as a health-systems input and his or her diminishment as a bearer of rights and an object of beneficence.

The roots of this redescription begin in the depths of all major cultures, all major religions, and every military tradition. Even its formal philosophical roots are long ones.

Writing to the King of Cyprus *On Princely Government*, Aquinas advised that "the good of the community is greater and more divine than the good of the individual. Thus the hurt of some individual is sometimes to be tolerated, if it makes for the good of the community; as, for instance, when a thief is put to death for the peace of the community. God himself would not permit evil in the world if good did not come of it."[4]

How much greater was the community good than the individual good? And how best might their magnitudes be computed? Hobbes took a step toward answering these questions: "The *Value*, or WORTH of a man, is as of all other things, his Price; that is to say, so much as would be given for the use of his Power: and therefore is not absolute; but a thing dependent on the need and judgment of another. . . . For let a man (as most men do,) rate themselves as the highest Value they can; yet their true Value is no more than it is esteemed by others."[5] Hobbes's market-clearing rule for human valuation specifically set aside the possibility that human value could in any way be absolute, that humans had value as humans. There is a great deal in *Leviathan* about the various qualities of man, good and bad, but nothing to suggest that any quality was external to a market mechanism. A man had value only as a tool, and only if a more powerful bidder, individual or corporate, wished to use him as a tool.

Along this Hobbesian line was Petty's "way of computing the value of every Head one with another." Petty figured that the cost of maintaining a male laborer or a country maidservant was about £7 a year.[6] There were some six million Britons assumed to be in the labor force in 1676. They paid about £8 million to rent land, and landowners spent another £8 million realized from the work performed on their estates. So, £42 million to maintain the working class minus £16 million in agricultural rent and produce had to leave £26 million in nonagricultural wages. Thus, the average man or woman earned "above 80 *l.* Sterling" during a twenty-year useful working lifetime; this was the individual's value to the nation. Petty discounted the value of the future work of children arbitrarily by half,[7] and he thought seamen were three times as valuable as husbandmen.[8] "[F]rom whence we may learn to compute the loss we have sustained by the Plague, by the Slaughter of Men in War, and by the sending them abroad into the Service of Foreign Princes."[9]

Petty's method was technically primitive, politically crude, and conceptually useful. It required the Hobbesian assumption that human value varied with market conditions, and it was untroubled by the incipient ideas that people of various classes were fundamentally interchangeable, that their futures by rights should have been self-determinable, that the sale (or rental) of their labor did not imply the sale of themselves, and that their labor, again by rights, *might* have won them something beyond their wages. The accounting methods traced to Petty have, with many refinements, survived, and their application in public and private affairs has itself become a major employer. They have even won a place in health-policy reform. Monetization of the value of human life, health, and happiness has long since lost its novelty in the health-policy literature, and discussions of particular diseases—influenza or arthritis or alcoholism or occupational injury rather than "the Plague"—are routinely (and often a bit ridiculously) prefaced by citation of social cost: so much time lost from work, so much money lost in treatment, so much advantage lost in international economic competition (but never "so much employment created in the domestic health-care sector").

How else might the social cost of a public-health problem be discovered? By measuring the number of people affected or the severity of the individual effect or some product of the two. Measuring the number of people affected suggests a uniformity, if not an equality, among individuals, while measuring the severity of the individual effect suggests the importance of unique predicaments. Measuring some product of number affected and severity of individual effect suggests the additivity of personal utilities and the *comparability* of compound utilities, and this comparability in turn suggests, as in Aquinas, that "the hurt of some individual is sometimes to be tolerated, if it makes for the good of the community."

This utilitarian calculus, into which Petty's "political arithmetick" can neatly be made to fit, is the work of many authors: Leibnitz, Smith, Ricardo, Bentham, Arsène-Jules-Emile Juvenal Dupuit (1804–1866), Alfred Marshall (1842–1924) and the economists of the "Cambridge School," and hundreds more. It has made possible the economic analysis of individual and group welfare, the interpersonal and intergroup comparison of utilities, the systematic comparison of the costs and benefits of specific policies, and the rank-ordering of a range of policies by their cost-benefit ratios. It is, to its advocates and admirers, the zenith of social objectivity[10] and the sharp edge of liberalism.

Consider contraception among cultural traditionalists in a poor country with a disastrously high birth rate *and* a predominantly heterosexual HIV transmission pattern. Should the publicly subsidized contraceptive method of first choice offer the lowest cost per birth avoided or per AIDS case prevented? Or should the expense of parallel programs or of a comprehensive and more sophisticated single alternative program be met by leaching funds from refugee feeding stations, childhood immunization efforts, a land-reform buy-out trust, a headwaters tree-planting project, a nuclear-weapons development laboratory, or a trade fair featuring value-added exports in whose manufacture comparative advantages are foreseen? A modern adviser might conscientiously decline to answer such questions without formal cost-benefit analysis or, failing that, a series of referenda.

The problem is an old one: serving the collective good while doing minimal violence to the individual good. The question is whether the individual should be protected by rights or by the best efforts of social engineers or by luck. Bentham objected strenuously to the first option, and his successors have characteristically wished for the first but endorsed the second, while some citizens of most societies still feel mainly reliant on the third.

Though unmistakably a Benthamite, William Stanley Jevons (1835–1882), who with Karl von Menger (1840–1921) and Léon Walras (1834–1910) formed the "first triumvirate" of neoclassical economics, compared and aggregated utilities infrequently, and then only with reluctance, caution, and disclaimer. In *The Theory of Political Economy* (1871), he avoided the practice altogether: "The reader will find, again, that there is never, in any single instance, an attempt made to compare the amount of feeling in one mind with that in another. I see no means by which such comparisons can be accomplished."[11] Jevonian utilitarianism was stubbornly, if not consistently, individualistic, and the social engineering that sprang from it remains ethically accessible.

Jevons thought medical charity often misdirected in metropolitan Britain, and he favored the attachment of a fee to many of its services, but his reasoning turned less on the virtue of "getting the prices right" in a free market than it did on the importance of getting the poor to play *some* role in the funding of a quasi-public good:

> I feel bound to . . . call into question the policy of the whole of our medical charities, including all free public infirmaries, dispensaries, hospitals, and a large part of the vast amount of private charity. What I mean is, that the whole of these charities nourish in the poorest classes a contented sense of dependence on the richer classes for those ordinary requirements of life which they ought to be led to provide for themselves. Medical assistance is probably the least objectionable of all the forms of charity, but it nevertheless may be objectionable. . . . Every hospital and free dispensary tends to relax the habits of providence, which ought to be most carefully cultivated, and which cannot be better urged than with regard to the contingency of sickness. *The Times* not long ago published some very remarkable and complete statistics . . . showing that the annual revenue of the established charities of London alone amounted to more than two millions a year. I fear that not only is a large part of this wasted in the excessive costs of management, but that a further large proportion really goes to undermine the most valuable qualities of self-reliance, and so form a bribe towards the habits of mendicancy and pauperism. . . . Now, I ask, why should the poorer classes be thus encouraged and instructed to look to the wealthier classes for aid in some of the commonest requirements of life? If they were absolutely unable to provide for themselves the reason would be a strong and intelligible one, but I do not believe that the people are really in such a hopeless state of poverty. On the contrary, the wages of the greater part of the working-classes, and in these districts [presumably Manchester and environs] almost the whole, are probably capable, if wisely expended, of meeting the ordinary evils and contingencies of life, and were providence in small matters the rule, the most unhesitating aid might properly be given in the more unforeseen and severe accidents and cases of destitution.
>
> But there is little use in bewailing an evil unless some mode of remedying it can be found. There is not much difficulty in discovering the only remedy applicable to medical charities. No one can seriously think of abolishing those charities; but why should not the working-classes be required to contribute towards institutions mainly established for their benefit? Self-supporting dispensaries exist in many places which afford all requisite aid to any person subscribing some such small amount as 1*d.* or 2*d.* each per week. I have heard that some of the London hospitals have considered the idea of adopting this system, and refusing aid in all minor cases but to their own subscribers. It would not be necessary to render the hospitals self-support-

ing. Endowments and public contributions would usually enable every hospital or dispensary to give back in medical aid several times the value of what is given in small contributions. The object would be not so much to raise money as to avoid undermining the prudent habits of the people. Non-contributors might still be relieved, but only on the payment of a fine; and, of course, cases of severe accident, illness, or destitution would still be relieved gratuitously as at present.[12]

One must not accept too readily the implication that being poor and sick a century ago in England was a warm and edifying experience. It must have been pretty miserable, even by standards of the day. The instructive aspect of Jevons's view is not that men and women and children had a right to health care and not even that they had a right to charity, but that health care *had to be provided,* even if through charity. The cost of routine health care for the poor was worthy of at least symbolic support from its beneficiaries, among whom savings could justifiably be enforced, and the cost of nonroutine or comparatively expensive or irresponsibly unsaved-for health care was worthy of support through the still-voluntary redistribution of funds from upper and middle to middle and lower classes. There was no suggestion that the health care any individual truly needed should go unsupported by society, though it must frequently have done so in practice (even in Manchester).

Jevonian utilitarianism was parentalistic; it was concerned with understanding and overtly with improving individual behavior, through a detailed sense of which social engineering would become more effective and more humane. Jevons drowned while on holiday near Hastings with his family, and utilitarianism, following Marshall's seldom challenged direction, took a more "utilitarian" course, analytically ambitious, ethically remote. In full modern array, it can help us lead ourselves to a cleverly chosen happiness we may not know we want and may not long respect.

Placing health care on a list of human rights[13] does not put it on a list of civil rights or on a list of social entitlements, and status as a human right does not mean that a particular good will be valued, much less guaranteed, while the greatest good for the greatest number is being identified, achieved, and accounted. Societies decide, idiosyncratically, whether health care is to be only a good or also a gift or, simply, a right. They have rarely if ever chosen or persisted in choosing the free-market option exclusively, compensatory charity being so powerfully subversive. They cannot choose charity alone, since those who would give to others will regularly give first to themselves and, to do so, will buy what they want when charity or a primitive mutual-aid arrangement fails to provide it. Societies ordinarily now choose to describe health care as a right, though sometimes insincerely and many times to ambiguous effect, their choice being, literally, *to describe*, not to establish, preserve, protect, or defend.

Some have argued that the existence of a *natural* right to health care is dubious and that the "creation" of such a right would infringe greater values, such as property rights, economic liberty, individual freedom, and the autonomy of communities.

To many, including some of the philosophers identified with the leading modern liberal school, health care is best valued not *per se* but as a leveler of opportunities, as a guarantor of the fair start and the fair shot: good health, then, for healthy competition.[14] While procedurally appealing in an era accustomed to arguments for electoral rights and economic or "marketplace" rights, embedding health care in a "fair-access" objective function gives it a factorial role incidental to—or even hostile to—clinical ethics. To what "opportunity" does health care provide "fair access" for the persistently insane or the profoundly retarded or for the terminally ill or terminally old? None in any plain sense. Could "opportunity" and "fair access" somehow be construed to fit such cases? Yes, but only with elaborately tendentious bending. Might patients lacking all "opportunity," conventionally defined, be thought to have even less claim to a "right to health care" than patients with better prognoses? Yes, as we have seen *and*, in some dressed-up way, could see again.

One modern philosopher, a libertarian, has compared society's responsibility to ensure access to health care to its responsibility to ensure access to haircuts[15] and has concluded, generally, "that the state may not use its coercive apparatus for the purpose of getting some citizens to aid others,"[16] including within this procedural proscription the "[t]axation of earnings from labor."[17] Another modern philosopher, a physician by education, has implied much the same: "People . . . have the secular moral right, no matter how unfeeling and uncharitable such actions may appear to others, not to aid those with excessive burdens, even if the financial burdens of those who could be taxed would not be excessive."[18] Correspondingly, he has concluded that "[a] basic human right to the delivery of health care, even to the delivery of a decent minimum of health care, does not exist."[19] A marketplace morality and a nonessentiality of "rights" have for him proved ultimately attractive:

> We must distinguish between positive and negative rights to health care. There are negative rights to health care as rights to be unobstructed in free associations with others in the marketplace. One should be able without hindrance to trade one's private resources for the services of free men and women. Positive rights to health care, rights to the delivery of health care, arise only insofar as they are created by particular communities or nations.

> There is no basic human right to health care or a fundamental human obligation to provide health care, though there may be very important reasons to create such a right or to see the obligation of being beneficent as including such a duty. . . .
>
> The character of rights to health care delivery will depend on the moral vision of a particular community or the result of the negotiations among moral communities as occurs when a nation fashions its health care system.[20]

It has also been argued prominently and passionately by one academic physician that the protection of a right to health care would itself be unethical:

> From man's primary right—the right to his own life—derive all others, including the rights to select and pursue his own values, and to dispose of these values, once gained, without coercion. The choice of the conditions under which a physician's services are rendered belongs to the physician as a consequence of his right to support his own life.
>
> If medical care, which includes physician's services, is considered the right of the patient, that right should properly be protected by government law. Since the ultimate authority of all law is force of arms, the physician's professional judgment—that is, his mind—is controlled through threat of violence by the state. Force is the antithesis of mind, and man cannot survive qua man without the free use of his mind. Thus, since the concept of medical care as the right of the patient entails the use or threat of violence against physicians, that concept is anti-mind—therefore, anti-life, and, therefore, immoral.[21]

One might have asked, as Schweitzer surely would have, whether denying a right to health care would not likewise have been "anti-life, and, therefore, immoral," but such a question might have served no cause save consistency.

The free-market romantic and libertarian ethicist are all for autonomy, and they are all for property, for its inheritance in whole and for its disposition in self-interest. And they all, one way or another, scratch their theories against the Lockean touchstone, proud to claim its proof of purity. Yet, were they to meet Locke, they might find he had a complaint, since they seem not to have read or liked much of his *First Treatise* and to have liked only the safer parts of his *Second*, for they have shunned his arguments for a right to charity and for a right to the redistribution of surplus. Life scientists, taken together, are responsible for the maintenance and increase of a sizable portion of the common inheritance of humankind. When significant nonrandom distinctions in access to their services—to their "charity" and their "surplus"—are knowingly left unremediated, the commonality of that inheritance becomes, by degrees, a fraud.

Theoretical provenance aside, free-market romantics and libertarian ethicists have little to show for their premise that health care is a good like any other and that those who bring it to market bear no specially altruistic responsibility. Allowing the guiding hand of self-interest to distribute *all* the comforts of scientific medicine and surgery is to allow what no modern peacetime society so far has had eyes cold enough to watch.

Antithetical to the libertarian ideal is an assertive communitarian notion that societies could "conscript" physicians, just as they have in earlier eras or under extraordinary conditions and to variable degrees "conscripted" priests, soldiers, teachers, and lawyers:

> I see no reason to respect the doctor's market freedom. Needed goods are not commodities. Or, more precisely, they can be bought and sold only insofar as they are available above and beyond whatever level of provision is fixed by democratic decision making (and only insofar as the buying and selling doesn't distort distributions below that level). . . .
>
> So long as communal funds are spent, as they currently are, to finance research, build hospitals, and pay the fees of doctors in private practice, the services that these expenditures underwrite must be equally available to all citizens.
>
> This, then, is the argument for an expanded welfare state. . . .
>
> Hence the three principles, which can be summed up in a revised version of Marx's famous dictum: From each according to his ability (or his resources); to each according to his socially recognized needs. This, I think, is the deepest meaning of the social contract. It only remains to work out the details—but in everyday life, the details are everything.[22]

Communitarian opinions such as this one respect the moral authority of customary societal groups and democratic majorities or pluralities with a nearly Rousseauian readiness that the life-sciences liberal would be unable consistently to share. The objections would *not* be libertarian; the life-sciences professional *does* have high social debts and extraordinary communal obligations. No, the objections would be *anti-majoritarian* and *anti-utilitarian.* Where is the patient's right to that which his society may not be pleased spontaneously to grant?

From the patient's point of view, libertarianism and communitarianism can look a lot alike. Many classes of patients may simultaneously lack the means to secure free-market health care and the influence or charm to secure social entitlement, formal or informal. They lie wounded in a no-man's land between market and charity, between good and gift. Social engineers have long tried to reach them, none more bravely than the economist.

Consider health policy and its reform in societies conspicuous for material fortune and civic sophistication. There are four stories here. One relates the genuine progress of high intention, and some of this story we have already told. A second story, now likewise familiar, is cynical inside: public health for national strength or for racial or ethnic vigor, for peace among the masses, for class or professional privilege. A third story combines the high intention of the first with the contrition of the second, the cautionary classic being the eugenics movement. A fourth story is again filled with high intention, now of the scholastic sort, and describes so strong an attachment to impartiality, so firm a bond to method, that individuality dims and charity becomes an error. It is to this fourth story that we shall now turn.

Denying a Right to Health Care

If health care is not a right, then access to health care can rightly be denied. How has it been denied in rich societies, and how has its denial been justified when and where political choice has explained "the scarcity of resources" more credibly than "the scarcity of resources" has explained political choice?

Now and then, the denial of access to health care has been the expected and intended and frankly desired direct or indirect effect of fully rational planning or expertly applied political pressure. Thus, the bearers of specified diseases[23] and the poor or primitive generally may not cross certain international borders to seek medical aid. More often, the denial of access to health care has been unintentional, even accidental, the product of chance or social circumstance or societal custom, the unexpected and undesired side effect of less-than-rational planning. Thus, the domestic poor or members of historically unfavored groups, such as racial minorities or women, may have found themselves unlikely candidates for expensive diagnostic procedures and complex treatments, despite the adequacy of public assistance or private means.[24] Other times, the denial of access to health care has been the altogether expected yet still undesired side effect of *fully* rational planning. Thus, societies determined to restrict the proportion of resources committed to health care have demonstrated that citizens without influence are likely to become citizens without benefits.

The most egregious rationalized denials of access to health care need not distract us here. We have already considered them by category and discussed a few of them by consequence. It is in the *less* egregious cases of rationalized denial, where human rights are rhetorically respected but health rights are boldly disputed, that ethics more easily gets bewildered.

In the United States, the care of poor pregnant women, poor mothers, and their inevitably and "innocently" poor children has been a "good" of unsettled value. The Maternity and Infancy Act of 1921—the Sheppard-Towner Act—was the first great achievement of universal suffrage. Through federal matching grants to the states, many useful services were provided, some directly life-enhancing, others indirectly life-saving, but no medical or surgical treatment could be supported. The American Medical Association argued effectively that such support would constitute "socialized" medicine and would by slow but certain course establish "socialism" in America. With economic and professional prerogative said to be in jeopardy, organized medicine opposed the guarantee of access to maternal-and-child health care, as did atavistic parochialists of many American stripes: nativists, states-righters, white-supremacists, free-marketeers, religionists, the unobliged "nobility," and old enemies of the vote for women. When it came up for renewal in 1927, Sheppard-Towner was funded only till 1929, to the disadvantage of millions who were poor and many millions more who soon would be. Not before 1935, the sixth year of worldwide economic emergency, would revival be politically possible, this time in the guise of Title V of the Social Security Act. Aid to "dependent children" was now forthcoming, and some crippling conditions could actually be treated; the president at the time, Franklin Delano Roosevelt (1882–1945), had never been poor, but he had been paraparetic since contracting poliomyelitis in 1921. The American Medical Association had to withdraw on the narrow front of aid to crippled children, who were stubbornly sympathetic when filmed with the president in his Warm Springs pool, but it was able to defeat broader advances. During World War II, the United States Children's Bureau was allowed to run the Emergency Maternity and Infant Care Program for the *wives* and legitimate children of servicemen of the lowest grades, but activities were reduced when peace came and ended entirely in 1949,[25] just when, in the United Kingdom, class-ridden and bankrupt and years from the end of "wartime" food rationing, the National Health Service was being formed on egalitarian lines.

In the United States, a determined effort to entitle at least the elderly to medical aid—"Medicare"—began in 1951. Thirteen years later, it succeeded and even entrained in its wake a widely variable, federally assisted but state-determined entitlement—"Medicaid" in Title XIX of the Social Security Act—for some of the poor, including some women and some children, for the elderly needing long-term care, for the blind, and for some of the disabled. The terms of these two entitlements, Medicare and Medicaid, changed many times in subsequent decades, and Medicaid particularly became an object of contention, criticized by poor and rich, patient and "provider," as the physician, the nurse, the hospital, and others—too rarely the midwife—collectively came to be called.

Despite its cost, which grew great indeed, the Medicaid quilt was too small and too thin. True, many needy people had irresponsibly neglected their future, and many neglected even their present. That said, millions of nonelderly and nonrich Americans remained entirely uninsured for health care, and the poor and misfortunate of many states had obvious needs Medicaid could not help them meet. These people were markedly heterogeneous and predictably unorganized if imagined as a group; many were nonvoters and, on the political balance scales, they were regularly outweighed by moneyed interests and consumer constituencies of even modest heft.

Could they be imagined *as a group* not just for public-accounting purposes but also for political and ethical purposes? Could they be told, as a group, "this much and no more" or "this much now and even less later"? Could they be told that their so-called right to health care had been explicitly amended and selectively revoked *in their own best interest* by a voting majority of their generally older, possibly healthier, mostly wealthier, and arguably wiser neighbors? Yes.

Hoping to save enough money to expand the number of less expensive benefits delivered, the Medicaid program in the state of Oregon—among the less generous of America's Medicaid programs—was required in 1987 by vote of the state's legislature to refuse some requests for high-cost procedures and treatments. Prior authorization for the transplantation of organs was to be denied in specified situations,[26] just as in several other states. Some patients in need left home for states where Medicaid rules were comparatively lenient.[27] Some patients tried to meet the market price of needed procedures through the generosity of family, friends, townsfolk, and television viewers or through the fundraising efforts of sympathetic celebrities or through the gift of funds "orphaned" by the premature demise of a fellow supplicant.[28]

To sustain such a policy against public discomfiture, the case for Medicaid restriction had to be made more effectively, and the advantages had to be more widely shared. In 1989, to distribute health care more equitably among Medicaid-eligible citizens *and to maximize the number of poor citizens served by Medicaid,* the Oregon legislature, advised by a priest-bioethicist and led by a physician-legislator, placed among the mostly progressive requirements of the Oregon Basic Health Services Act and related acts and amendments a particularly audacious innovation: a plan to rank *all* medical conditions and *all* medical and surgical treatments—all "condition-treatment pairs"—by a measure of worthiness[29] and then to provide to *all* previously uninsured nonelderly poor patients a deliberately shortened list of benefits. All certifiably poor people, including all poor women and children, would at last be insured, but those *already* insured would lose benefits to the newly insured, typically adult males; losers might lose entirely the very benefits they needed while gaining others they could never use, as might their neighbors, with whom they could not make trades. By electoral chance, this multistep experiment in utilitarian chemistry was about to be run with less money and more heat than expected.

State governments in the United States have usually taxed property in addition to or instead of income or consumption. Though conceptually progressive, taxing property can become distinctly unpopular as property values, local demography, and secular economic conditions change. Such was the case in Oregon in the 1980s, and, though they taxed themselves overall at only 90.3 percent of the United States average, Oregonians taxed their property at 140 percent of the average. The result was "Ballot Measure 5," upon which those Oregonians eligible to vote, registered to vote, and choosing to vote voted approvingly, 52 percent to 48 percent, in November 1990.

"Ballot Measure 5" required increasingly drastic cuts in property taxes over a five-year period. So, it required at least one of three responses: the raising of replacement revenues or the elimination of "waste, fraud, and abuse" or the increasingly drastic reduction of publicly funded services. The governor initiated what she called "A Conversation with Oregon," a nine-month community-meeting process designed to help citizens clarify and, in a quasi-democratic manner, enunciate their own expectations. A major theme was the ranking of Medicaid's new "condition-treatment pairs" by their "importance."

Participating in the community-meetings process were something over 1,000 Oregonians; of these, 90.6 percent were insured and 9.4 percent were uninsured; only 4.4 percent were insured by Medicaid. Poor women and children, patients in long-term care, and patients whose illnesses had impoverished them—patients who had "spent-down to Medicaid"—were not well represented in the "Conversation." They or their surrogates could have been, but, on the whole, they were not; anyway, their views on the relative importance of hundreds of "condition-treatment pairs" probably resembled that of their fellow citizens.[30] For better or for worse, about two-thirds of participating citizens were employed in the health-care sector.[31]

An eleven-member commission of physicians and others drew on this community-meeting experience and on the results of telephone surveys and expert interviews and on more ordinary sources, such as personal judgment and professional literature, to rank 709 (or 714) "condition-treatment pairs"; revenue projections suggested that the most important 587 would be available to Medicaid beneficiaries in the plan's first year, fewer in future years. The ranking process, start to finish, was difficult and fascinating, and from its study much could be said about popular and professional attitudes toward disease and its treatment, about the reproducibility of preference measures, about the "effectiveness" of cost-effectiveness analysis, which evolved from formality to informality to virtual irrelevance, and about committee method and social conscience.[32]

The commission was trying "to compare the amount of feeling in one mind with that in another," as Jevons would have put it, and, despite revisions of method *and* revisions of results, its product was hard to accept as a final working rule. Medicaid would buy prenatal care for *all* poor women and it would buy the management of hypertension and diabetes and the treatment of angina pectoris and acute conjunctivitis and acne and the surgical removal of an inflamed appendix and the medical therapy of *cervical* intervertebral disk disease and the transplantation of kidneys and corneas. But Medicaid would not buy cosmetic surgery for mastectomy patients or the transplantation of bone marrow or heart or pancreas or the transplantation of livers into alcoholics or the medical therapy of *thoracolumbar* intervertebral disk disease or intensive care for AIDS patients likely to live no more than six months if saved immediately. It would buy any attempt at cardiopulmonary resuscitation, no matter how futile, but it would buy no treatment for cancer in a patient whose five-year treated survival chance was less than 10 percent, even if palliative therapy offered excellent pain relief (e.g., radiation for the bone metastases of breast cancer) or months—or several years—of near-normality (e.g., castration for metastatic adenocarcinoma of the prostate). Medicaid would buy any amount of most procedures but none of others, and if revenues fell short in the first year or *when* revenues decreased in subsequent budget cycles, the list of goods and services Medicaid would purchase would shorten by a few or by several dozen or by hundreds of "condition-treatment pairs." Among the first procedures to be dropped would be repair of a congenitally dislocated hip (whose neglect can be crippling), medical and surgical treatment of spontaneous or missed abortion (whose neglect can be fatal), and removal of a foreign body from the eye, usually among the simplest and always among the most appreciated of primary-care procedures.

In the near and leaner future, could a Medicaid patient with a splinter in his cornea have sued a doctor or a hospital for refusing to treat him? No; refusal to administer an uninsured treatment to a Medicaid patient had been immunized against claims of liability in state court.[33] The Oregon plan in this respect reified the legislature's judgment that property rights—here, the right to a fee for service—superseded professional and corporate obligation in health care as in other types of commerce. Could a Medicaid patient with a splinter in his cornea but with a surgically absent appendix have traded his access to a future appendectomy for cheap and immediate mercy toward his eye? No; all utilities were aggregate utilities, and all patients were average patients. The Oregon plan in this respect resembled the old Soviet economy: all local needs anticipated by the "center," all rationalizing exchanges among factory managers and among individual workers disallowed. Could a physician or a hospital profitably have accommodated a patient with a corneal foreign body by using Medicaid "savings" to cover Medicaid "costs"? No; reimbursement was not to be *per capita* but *per curationem.* The Oregon plan in this respect was driven by yes-or-no diagnosis, not by individual need or individual likelihood of benefit and, so, not by unique predicament, though clinical practice is *nothing other than* the evaluation and management of an endless series of unique predicaments—a truth unchanged by resource constraint, a truth unchanged even by war. Treating the sickest first and the more treatable before the less treatable—jointly, the best habits of triage—would have better served individual need, but it would have exponentially complicated the ranking system and its consequent reimbursement rules, and, depending on the level and the "school" of economic analysis, it still might not have been "surplus-enlarging" or "welfare improving."

Interest groups steamed in review past a congressional hearing in September 1991. Predictably *against* the plan were, among others, the Children's Defense Fund, Families USA, the National Association of Children's Hospitals and Related Institutions, the National Council of Senior Citizens, the United Cerebral Palsy Associations, the United States Catholic Conference. Predictably *for* the plan were, among others, the Oregon affiliate of the American Diabetes Association, Associated Oregon Industries, the Business Group on Health, Legacy Health System ("a not-for-profit system of healthcare services"), the Oregon AFL-CIO, the Oregon Association of Hospitals, the Oregon Department of Human Resources, Oregon Health Decisions ("a citizen network for education and action on ethical issues in health care"), the Oregon Health Sciences University (a major provider of uncompensated care to the poor), the Oregon Medical Association, and the Sisters of Providence in Oregon and the Sisters of St. Joseph of Peace (operators of private sectarian not-for-profit hospitals and health-care plans).[34]

Even (or especially) among friends of the Oregon Basic Health Care Plan, the need for further tinkering was acknowledged, but, as far as the Bush administration (1989–1993) was concerned, tinkering would not do.[35] Oregon's plan required the suspension of federal Medicaid rules,[36] and the federal government did not agree to suspend them, pointing out that the recently signed Americans With Disabilities Act prohibited discrimination by condition for at least some citizens.[37] This federal decision was thought by some to be principled and providential,[38] but others regarded it variously as obstructionistic, shortsighted, small-minded, counterproductive, disingenuous, chilling, rude, and cowardly.[39] Writing opposite the editorial page in the *New York Times*, the governor of Oregon complained that "the Administration put its courage in the closet when it rejected Oregon's proposal":

> Our plan is an honest answer to the nation's health care problem. It recognizes that no state can afford to provide every possible medical service and that choices must be made. Where other states are able to make medical care available to only a fraction of those in poverty, the Oregon plan would make the most effective care available to *everyone*.
>
> The plan would have expanded Medicaid to cover all Oregonians below the poverty level. To do this, Oregon would have eliminated coverage of certain services medical practice shows to be less effective than others.
>
> No element of the plan has received more scrutiny than this ranking of treatments. A list of all medical treatments was drawn up . . . [that] emphasized preventive and primary care and maternal and child services. . . .
>
> The Department of Health and Human Services has had our waiver application for almost a year. We worked with Federal officials at every step to resolve all their concerns. But the only time we got direct word that they believed our plan conflicted with the Americans With Disabilities Act was in the letter of denial. Where were those concerns for the last year? Why wasn't Oregon been given a chance to respond?
>
> My husband is a paraplegic and one of my sons is autistic. I have spent my life working as an advocate for people with disabilities. As Governor, I would not have supported a plan that compromised their rights and needs. Oregon's plan chooses effective treatments over ineffective ones. The administration claims that such choices discriminate against people with disabilities. Here is why they are wrong:
>
> • The plan does not apply to the aged, blind and disabled until 1993, when the priority list would be reviewed to insure it meets the needs of those groups. Until then, disabled people who are able would continue to receive Medicaid. . . .
>
> • Oregon's plan would offer high-quality health care to many poor, disabled people who now have no health care at all because they aren't eligible for Medicaid. . . .
>
> [T]he Bush Administration has expressed concern that the plan violates the "Baby Doe" laws that protect extremely premature babies. That's plain wrong. The food, water and medication prescribed by those laws are all covered under the plan. . . .
>
> How long do we and the rest of the nation have to wait for a fair, rational system of health care?[40]

The editor boldly superimposed in the middle of this essay a question for the browsing breakfast reader: "If not the Oregon plan, then what?" Members-to-be of the Clinton administration (inaugurated 1993) may have been storing up the answer; once in office, they revivified the application, again revised.[41]

The notion that some health-care goods and services were more valuable than others was neither new nor exotic. It followed that patients needing a range of health-care goods and services could rationally and ethically be left untreated if conditions so demanded. Did conditions in Oregon demand such a practice? Of course not. Oregon was not at war. Its economy had not collapsed. It saw no reason—and had no reliable mechanism by which—to redistribute health-care goods and services from areas of "excess" supply to areas of "excess" demand. Oregon might have realized comparable or greater savings by refusing to subsidize the maintenance of unneeded hospital beds or by discouraging unindicated hospital admissions for a range of common conditions.[42] It could have refused payment for unnecessary diagnostic procedures and for expensive newer medicines offering no significant advantage over inexpensive older medicines. It could have looked "overseas" to an often-visited but rarely studied fellow state: Hawaii, where a patched-together nearly universal insurance system was generating best-in-the-nation public-health results at or near lowest-in-the-nation per capita costs.[43] Or it might have looked abroad to any federal or provincial or regional plan likewise spending less and getting more. Or it could have applied to otherwise-unfunded health care (not to gymnasia and the timber industry) the proceeds of its newly established state-run lottery, which cleared over $600 million between 1985 and 1991 through controversial games based on the outcomes of professional athletics contests and to which the addition of video poker and "keno" (which latter entertainment would prove highly profitable) was being considered.[44] Oregon could have taken hold of its systemic problem in a way that made affordable whatever was wisely requested, regardless of social class. Instead, "Oregon"—a personified concatenation of majorities and pluralities—chose to optimize its social-welfare function in a Benthamite manner, supposedly fair, presumably efficient, potentially happiness-increasing, and necessarily unrestricted by any individual right to health care.

Why "necessarily"? Because communal preference is sanctified in utilitarian ethics, and the sanctity of communal preference is incompatible with non-communal rights. The Oregon plan required that *in the matter of health-care for the poor* the community would have rights and the individual would have privileges. Rights, of course, are not privileges, not "private laws"; rights are not subject to legislative extension and withdrawal, but only to recognition or denial, protection or violation. Not that the voting people of Oregon would have pressed for tax relief requiring—or would *ever* have allowed, regardless of incentive—the Stalinesque dispatch of their social and economic "enemies." In American political culture, any such initiative would have been the mark of mass insanity, and, in any event, the United States Constitution *as amended* would have barred its way. But the right to health care was not held to be self-evident in the United States; it was not thought to be inalienable, and its recognition was not a prior constraint on health-care reform. The right to health care was rhetorical only.

Imagine the right to trial by jury in felony cases being rhetorical only. Voters might decide that jurors should be paid market-price consulting fees by the defendants demanding their attention and that the public payment of jurors should somehow be limited according to type of accusation and prior probability of guilt. In simple and easily predicted cases involving poor defendants, summary judgment might suffice. In simple but hard-to-predict cases involving poor defendants, cheap jurors might be recruited from the court-house steps or from nearby street corners or from the waiting rooms of for-profit blood banks. In complex cases involving poor defendants, jury trial would require clever and sophisticated citizens whose high fees would have to be funded by the state through taxation of property or income or commerce; summary judgment here might actively be required, since an occasional false verdict would be a misfortune on the scales of public happiness far lighter than a property tax or a general levy.

If Oregon had recognized a right to health care in its constitution or if the Oregon legislature had decided *effectively* to recognize such a right, despite constitutional silence, health-care reform would have assumed a different shape—as it had in the Canadian provinces a generation before.[45] That to which a patient had a right would still have depended on societal capacities and indirectly on democratically expressed communal preferences, some wise and some not. If a treatment or procedure was to be unavailable, no right would secure it; if it *was* to be available, no right to it could unequitably be denied. But the right recognized in Oregon in the late 1980s and early 1990s was a different one, communal and majoritarian.

Locke in the *Essay* argued against the innateness of ideas. He knew that a right to charity, however natural, could not be discovered preformed in the mind. Locke knew that the right to charity had to be "learned" in ethics and had to be *declared* in politics. And he knew that, once declared, it would still perpetually have to be defended, not so much against evil as against competing goods. Inviting the medically indigent to graze the health-care commons as if they were co-equal contributors to its maintenance is particularly hard to defend, for it seems to diminish fairness, undermine order, confiscate property, and indulge sloth. Many poor and sick adults and the poor parents of many sick children fail in the civic virtues or fail so fully in their luck that assumptions of their worthiness wear out. Generosity stretches and sags with the insistence, repetitiveness, and chronicity of their need, whether or not the product of intemperance. All true. But, in any year and in any place, even at death's door or on death row, does the *least* admirable penniless patient deserve the removal of a foreign body from the eye? If the answer is yes, then the right to health care is real.

Could the right to health care be as "real" as any human right and yet be bartered for another good, as when the liberty or the life of a felon is "paid" to cancel a debt incurred in crime? No. The patient is not a criminal; even the criminal who is sick is not a criminal patient. Forfeiture of the right to health care is punitive *per se* and cannot rightly be a feature of any noncapital punishment, even lifelong incarceration—hence, the prohibition of torture in all forms and the special criminality of physician participation in execution, as denounced first by Celsus, later by Bernard, and not last by the Nuremberg Tribunal. Forfeiture of the right to health care is prerequisite for capital punishment and reclassifies *as capital* any punishment called by a nicer name and reclassifies *as punishment* any arbitrary denial.

Could the right to health care be as "real" as any human right and yet be securable only through self-help? Or could it be only a negative right and not a positive one, only a freedom from maliciousness, not an entitlement to beneficence? No. As early as Browne and Boyle and as vigorously as Virchow and Schweitzer, the duty of those educated to care for the health of their fellows has been determined and described, more or less safely, in a socially contractarian way: life-sciences research and medical practice must be taken up *much* less selfishly and more activistically than other learned professions. Schweitzer added a greater duty, universal, non-contractarian, and paraprofessional: while all intelligent beings should honor all else that lives, those who best understand life should be expected most faithfully to revere it.

Could the right to health care be as "real" as any human right and yet the "right" of one patient be bartered for the individually less expensive "rights" of a collection of other patients or for some other manifestly popular collective good? No, not in a rich country at peace. Society *alone* would there and then have become the rights-bearer, with health care a privilege enjoyed at societal discretion by subsets of citizens. As in spirit since Scribonius and as in print since Percival, a law enacting such a scheme would be a law worth violating.

XIII

GROPING IN THE LIGHT

HOMINES SAPIENTES are smart enough to doubt their own moral wisdom. And, when they feel the need to revise it, they are timid enough to seek permission. Even scientists, who must learn the skills of irreverence, seek authority in ethics. Even scientists know that wisdom has a sedimentary geology and cannot be understood from its surface.

Yet, most sediment is rubbish, and the case *against* ethical tradition, the case *against* valuing history more than immediacy, is strong. Tradition is the enemy of progress, and progress in its principal dimension is advantageous by definition. In ethics, however, progress is hard to define. Tradition, usually to its credit, is likewise the enemy of ethical revolution, and, mutations that they are, ethical revolutions must more often than not be maladaptive. Many are attempted; few attempts are noticed; fewer still succeed; and just a handful are remembered admiringly.

We may have uncovered a serviceable past, but we knew what we were looking for, and we cast aside much debris (and some toxic waste) to find it. Who is to say we valued the best and rejected the worst? What we have found we will call by its new name, life-sciences liberalism, but we will not call it "truth."

How would this old-and-new "ism" have its followers proceed? Where would it have them go, and what would they do and not do when they got there?

First, they would travel in midday. Their method would be historical and plain. Their judgment would draw upon any available experience, remote or recent, philosophical or scientific, but they would recognize no authority beyond good reasoning. Ethical arguments based *necessarily* on scientifically disproven world views would not be respected, though the *people* making such arguments and holding such views would be. No special credence would be given to the modern or even to the natural. Evolutionary sociobiology would be studied intensively for its explanatory power but would never be asked (or used) to tell right from wrong.

With civility but not apology, they would argue *and they would act* for the protection of individual human rights, including, idiosyncratically, the right to health care, realizing that students and masters of the life sciences have inherited an unusually heavy responsibility in this regard and that they must often on principle play an anti-majoritarian role—

one played happily *but only coincidentally* for the greater good. They would for some of the same reasons—not by extension from humanitarianism *but for some of the same reasons*—do their best to treat and have others treat nonhuman animals compassionately, even though these animals might be used (and eaten) against their animate and evident wills. They would also dissent from the brazen spoilage of nonanimal life, once again arguing less against human irrationality than *for* natural right.

They would accept that human nature makes us all, more or less, the product of our nurturing, and they would see their mirrored selves in domestic antagonists and foreign enemies. They would refuse, rudely or illegally if necessary, to use their skill as a punishment or as a weapon, and they would never allow themselves to enter their action anonymously as a factor in any equation of nationalistic predation, corporate cynicism, or societal arrogation. Nor would they deny their mercy to the few or the poor or the young when tempted or instructed to do so by the many or the rich or the "wise."

And they would never forget on whose prior efforts and ongoing sacrifices their own successes have had to depend.

Life-sciences liberalism is a discovery, not an invention. It was never intentionally hidden and has never entirely been out of use, yet it long lay somehow unseen in a philosophical gloom. There must be more still to find, and there is surely more to make, and some of what now seems to mean this or that may on maturing reflection be seen to mean the other. Its better judgment and fuller application are left respectfully and affectionately to the reader.

NOTES

CHAPTER I
A HISTORY OF CONVICTIONS

1. Leo Szilard, "My Trial as a War Criminal," *University of Chicago Law Review* 17, no. 1 (Autumn 1949): 79–86.

2. *Ibidem*, at asterisked note.

3. *Ibidem.*

4. Ezekiel J. Emanuel, *The Ends of Human Life: Medical Ethics in a Liberal Polity* (Cambridge, Mass.: Harvard University Press, 1991), pp. 9–41 at p. 29.

5. James D. Watson and John Tooze, *The DNA Story: A Documentary History of Gene Cloning* (San Francisco: W. H. Freeman, 1981), pp. 24–49.

6. Bernard Lown, "Looking Back, Seeing Ahead," *Lancet* 332, no. 8604 (July 23, 1988): 203–204.

7. V. W. Sidel, J. Geiger, and B. Lown, "The Physician's Role in the Post-Attack Period," *New England Journal of Medicine* 266 (1962): 12–20.

8. Bernard Lown, "The Physician's Commitment," in *The Final Epidemic: Physicians and Scientists on Nuclear War*, Ruth Adams and Susan Cullen, eds. (Chicago: Educational Foundation for Nuclear Science, 1981), pp. 237–240.

9. Lown, "Looking Back, Seeing Ahead"; Jennifer Leaning, "Physicians, Triage, and Nuclear War," *Lancet*, no. 8605 (July 30, 1988): 269–270; Ian Maddocks, "Nuclear Threat and Health in the Pacific Ocean," *Lancet*, no. 8606 (August 6, 1988): 323–324; John E. Mack, "The Enemy System," *Lancet*, no. 8607 (August 13, 1988): 385–387; V. W. Sidel, "The Arms Race As a Threat to Health," *Lancet*, no. 8608 (August 20, 1988): 442–444; Michael McCally and Christine K. Cassel, "Nuclear Weapons Test Ban 1988," *Lancet*, no. 8609 (August 27, 1988): 495–496; Herbert L. Abrams, "Inadvertent Nuclear War," *Lancet*, no. 8610 (September 3, 1988): 559–560; William R. Beardslee, "Youth and the Threat of Nuclear War: The Psychological Task of Venturing into Unknown Territory," *Lancet*, no. 8611 (September 10, 1988): 618–620; Jeremy Leggett, "Could We Safely Negotiate a Treaty Banning All Nuclear Tests?" *Lancet*, no. 8612 (September 17, 1988): 674–675; Osvaldo Velasquez, "A View from a Nation Less Likely to Be a Target for Nuclear Weapons," *Lancet*, no. 8613 (September 24, 1988): 732–733; Thomas Piemonte, "Nuclear Winter," *Lancet*, no. 8614 (October 1, 1988): 785–786; Michael McCally, Christine Cassel, and Leah Norgrove, "Medical Education and Nuclear War," *Lancet*, no. 8615 (October 8, 1988): 834–837.

10. McCally et al., "Medical Education."

11. Carl N. Degler, *In Search of Human Nature: The Decline and Revival of Darwinism in American Social Thought* (New York: Oxford University Press, 1991), pp. 310–327 at p. 324.

12. Edward O. Wilson, *Sociobiology: The New Synthesis* (Cambridge, Mass.: Belknap Press of Harvard University Press, 1975), p. 575.

CHAPTER II
FROM FIRST PROBLEMS TO THE EDGE OF MODERNITY

1. Galen, *On the Natural Faculties*, Arthur John Brock, trans. (Cambridge, Mass.: Harvard University Press, 1963), book I, sec. xiii, pp. 49–71.

2. Hippocrates, *Ancient Medicine* in *Hippocrates*, W.H.S. Jones, trans. (Cambridge, Mass.: Harvard University Press, 1962), vol. 1, p. 53.

3. Ludwig Edelstein, "The Role of Eryximachus in Plato's *Symposium*," in *Ancient Medicine: Selected Papers of Ludwig Edelstein*, Owsei Temkin and C. Lilian Temkin, eds. (Baltimore: Johns Hopkins University Press, 1967, 1987), pp. 153–171 at pp. 165–171.

4. Aristotle, *Politics*, H. Rackham, ed. and trans. (Cambridge, Mass.: Harvard University Press, 1977), book 4, chapter 1, pp. 277–279.

5. Ludwig Edelstein, *Hippocrates the Oath or the Hippocratic Oath* (Chicago: Ares Publishers, 1979), pp. 15–48.

6. G.E.R. Lloyd, *Science and Morality in Greco-Roman Antiquity* (Cambridge, U.K.: Cambridge University Press, 1985), p. 5.

7. Celsus, *De Medicina*, W. G. Spencer, trans. (Cambridge, Mass.: Harvard University Press, 1960), vol. 1, pp. 13–15.

8. *Ibidem*, p. 15.

9. *Ibidem*, pp. 23–25.

10. *Ibidem*, p. 41.

11. Carleton B. Chapman, *Physicians, Law, and Ethics* (New York: New York University Press, 1984), pp. 40–41.

12. Scribonius Largus, *Compositiones*, Edmund D. Pellegrino and Alice A. Pellegrino, trans., in Edmund D. Pellegrino and Alice A. Pellegrino, "Humanism and Ethics in Roman Medicine: Translation and Commentary on a Text of Scribonius Largus," *Literature and Medicine* 7 (1988): 22–38.

13. Ludwig Edelstein, *Ancient Medicine: Selected Papers of Ludwig Edelstein*, Owsei Temkin and C. Lilian Temkin, eds. (Baltimore: Johns Hopkins University Press, 1967), pp. 339–340.

14. Scribonius Largus, *Compositiones.*

15. Edelstein, *Ancient Medicine*, p. 340.

16. Chapman, *Physicians, Law, and Ethics*, p. 43.

17. Galen, *On the Natural Faculties*, Arthur John Brock, trans. (London: William Heinemann, and Cambridge, Mass.: Harvard University Press, 1963), book 1, chapter 13, pp. 49–71.

18. William Osler, "The Old Humanities and the New Science: The Presidential Address Delivered Before the Classical Association at Oxford, May, 1919, by Sir William Osler, Bt., M.D., F.R.S., Regius Professor of Medicine, Oxford," *British Medical Journal*, July 5, 1919, pp. 1–7 at pp. 4–5.

19. Aristotle, *Politics*, book 1, chapter 1, pp. 3–7.

20. Lewis Hanke, *Aristotle and the American Indians: A Study in Race Prejudice in the Modern World* (Chicago: Henry Regnery, 1959), pp. 12–61 at pp. 54–61.

21. Lewis Hanke, *The Spanish Struggle for Justice in the Conquest of America* (Philadelphia: University of Pennsylvania Press, 1949), pp. 113–115.

22. Manuel Giménez Fernández, "Fray Bartolomé de Las Casas: A Biographi-

cal Sketch," in *Bartolomé de Las Casas in History: Toward an Understanding of the Man and His Work,* Juan Friede and Benjamin Keen, eds. (DeKalb: Northern Illinois University Press, 1971), pp. 67–125.

23. Bartolomeo de las Casas, *The Tears of the Indians: Being an Historical and true Account of the Cruel Massacres and Slaughters of above Twenty Millions of innocent People; Committed by the Spaniards in the Islands of Hispaniola, Cuba, Jamaica, etc. As also, in the Continent of Mexico, Peru, and other Places of the West Indies, To the total destruction of those Countries,* John Phillips, trans. (London, 1656), presented in facsimile as Bartolomeo de las Casas, *The Tears of the Indians* (Stanford, Calif.: Academic Reprints, undated).

24. Daniel J. Boorstin, *The Discoverers* (New York: Random House, 1983), pp. 626–635.

CHAPTER III
FROM THE SCIENTIFIC ATTITUDE TO THE UNIVERSALIST SENTIMENT

1. Francis Bacon, "Of Empire. XIX," in Francis Bacon, *The Essayes or Counsels, Civill and Morall,* Michael Kiernan, ed. (Cambridge, Mass.: Harvard University Press, 1985), pp. 60–61.

2. Francis Bacon, "Of the true Greatnesse of Kingdomes and *Estates.* XXIX," in Francis Bacon, *Essayes,* p. 95.

3. *Ibidem,* pp. 96–97.

4. Francis Bacon, "Of Seditions *And* Troubles. XV," in Francis Bacon, *Essayes,* p. 46.

5. Francis Bacon, "Of Unity in Religion. III," in Francis Bacon, *Essayes,* p. 14.

6. Francis Bacon, "An Advertisement Touching an Holy War," in Francis Bacon, *The Works of Francis Bacon, Baron of Verulam, Viscount St. Alban, and Lord High Chancellor of England,* James Spedding, Robert Leslie Ellis, and Douglas Denon Heath, eds. (London: Longman and Company, 1858), published in facsimile as Francis Bacon, *The Works of Francis Bacon* (Stuttgart-Bad Cannstatt: Friedrich Frommann Verlag, Guenther Holzboog, 1963), vol. 7, pp. 17–36.

7. *Ibidem,* vol. 7, p. 35.

8. Claude Bernard, *An Introduction to the Study of Experimental Medicine* (1865), Henry Copley Greene, trans. (New York: Macmillan, 1927), presented in facsimile as Claude Bernard, *An Introduction to the Study of Experimental Medicine* (Birmingham, Ala.: Classics of Medicine Library, 1980), p. 51.

9. Peter Urbach, *Francis Bacon's Philosophy of Science: An Account and a Reappraisal* (LaSalle, Ill.: Open Court Publishing Company, 1987), pp. 17–24.

10. *Ibidem,* pp. 187–193.

11. *Ibidem,* pp. 109–121.

12. Charles Webster, *The Great Instauration: Science, Medicine, and Reform, 1626–1660* (London: Duckworth, 1975), pp. 15–27.

13. Francis Bacon, *New Atlantis* in *Ideal Commonwealths,* rev. ed. Henry Morley, ed. (London: Colonial Press, 1901), pp. 103–137.

14. Robert Nisbet, *History of the Idea of Progress* (New York: Basic Books, 1980), p. 114.

15. Francis Bacon, *The Two Bookes of Francis Bacon of the Proficience and Advancement of Learning Divine and Humane* [*The Advancement of Learning*] (London: Henrie Tomes, 1605), in Francis Bacon, *The Works of Francis Bacon*, vol. 3, pp. 253–491 at p. 327.

16. Francis Bacon, "Religious Meditations," in Francis Bacon, *The Works of Francis Bacon*, vol. 7, pp. 243–254 at p. 253; Francis Bacon, *The Great Instauration [, Second Part]: The New Organon [or True Directions Concerning the Interpretation of Nature]*, in Francis Bacon, *The Works of Francis Bacon*, vol. 4, pp. 37–248 at pp. 120–121.

17. Urbach, *Francis Bacon's Philosophy of Science*, pp. 59–60.

18. Bacon, *The Great Instauration[, Second Part]*, vol. 4, p. 47.

19. Francis Bacon, "Of the Interpretation of Nature," in Francis Bacon, *The Works of Francis Bacon*, vol. 3, pp. 217–252 at pp. 222–223.

20. Webster, *The Great Instauration*, pp. 21–22.

21. Francis Bacon, "Of the Interpretation of Nature," vol. 3, pp. 217–252 at p. 217.

22. Francis Bacon, "Of Great Place. XI," in Francis Bacon, *Essayes*, p. 34.

23. Robert Nisbet, *History of the Idea of Progress* (New York: Basic Books), pp. 112–115.

24. Francis Bacon, "Of Goodnesse *And* Goodnesse of Nature. XIII," in Francis Bacon, *Essayes*, p. 39.

25. Francis Bacon, *The Great Instauration[, First Part]*, in Francis Bacon, *The Works of Francis Bacon*, vol. 4, pp. 3–36 at pp. 20–21.

26. Stephen Beasley Linnard Penrose, Jr., "The Reputation and Influence of Francis Bacon in the Seventeenth Century," Ph.D. diss., Columbia University, 1934, pp. vii–viii.

27. C. S. Lewis, *The Abolition of Man, or Reflections on Education with Special Reference to the Teaching of English in the Upper Forms of Schools* (New York: Macmillan, 1947), pp. 34–50 at pp. 35–37.

28. Leon R. Kass, "The New Biology: What Price Relieving Man's Estate? Efforts to eradicate human suffering raise difficult and profound questions of theory and practice," *Science* 174, no. 4011 (November 19, 1971): 779–788.

29. Nisbet, *History*, pp. 32–34.

30. Anthony Quinton, *Francis Bacon* (Oxford: Oxford University Press, 1980), pp. 70–71.

31. C. A. Patrides, " 'Above Atlas His Shoulders': An Introduction to Sir Thomas Browne," in Thomas Browne, *Sir Thomas Browne: The Major Works*, C. A. Patrides, ed. (Harmondsworth, U.K.: Penguin Books, 1977), p. 21.

32. Thomas Browne, *Religio Medici* in *Sir Thomas Browne: The Major Works*, pp. 140–141.

33. *Ibidem*, p. 133.

34. *Ibidem*, p. 134.

35. *Ibidem*, pp. 137–138.

36. Patrides, " 'Above Atlas His Shoulders,' " pp. 32–38.

37. Francis Bacon, *The Advancement of Learning* in Francis Bacon, *The Works of Francis Bacon*, vol. 3, pp. 253–491 at p. 364.

38. C. A. Patrides, "An Outline of Browne's Life within the Context of Contemporary Events," in Thomas Browne, *Sir Thomas Browne: The Major Works*, p. 18.

39. Patrides, "'Above Atlas His Shoulders,'" pp. 26–27.

40. Michael Wilding, "*Religio Medici* in the English Revolution," in *Approaches to Sir Thomas Browne: The Ann Arbor Tercentenary Lectures and Essays*, C. A. Patrides, ed. (Columbia: University of Missouri Press, 1982), pp. 100–114.

41. John T. Harwood, "Introduction," in *The Early Essays and Ethics of Robert Boyle*, John T. Harwood, ed. (Carbondale: Southern Illinois University Press, 1991), p. 1.

42. John Harrison and Peter Laslett, *The Library of John Locke*, 2d ed. (Oxford: Clarendon Press, 1971), p. 95.

43. Lord Edmond Fitzmaurice, *The Life of Sir William Petty, 1623–1687, One of the First Fellows of the Royal Society, Sometime Secretary to Henry Cromwell, Maker of the 'Down Survey' of Ireland, Author of 'Political Arithmetic' &c., Chiefly Derived from Private Documents Hitherto Unpublished* (London: John Murray, 1895).

44. Samuel Pepys, entry for January 27, 1664, in *The Diary of Samuel Pepys*, Robert C. Latham and William Matthews, eds. (Berkeley: University of California Press, 1971), vol. 5 (1664), p. 27.

45. William Osler, "Sir Thomas Browne," address to the Physical Society, Guy's Hospital, London, October 12, 1905, published in the *British Medical Journal*, 1905, ii, pp. 993–998.

46. Harvey Cushing, *The Life of Sir William Osler* (Oxford: Clarendon Press, 1925), published in facsimile as Harvey Cushing, *The Life of Sir William Osler* (Birmingham, Ala.: Classics of Medicine Library, 1982), vol. 1, pp. 501–502.

47. William Osler, "Bed-side Library for Medical Students," in *Aequanimitas, with Other Addresses to Medical Students, Nurses and Practitioners of Medicine*, 3d ed. (New York: McGraw-Hill, no date), endpaper (unnumbered).

48. Cushing, *The Life of Sir William Osler*, vol. 2, p. 686.

49. "Introduction," in *Religio Medici, Sir Thomas Browne: Notes from the Editors* (Birmingham, Ala.: Classics of Medicine Library, 1981), p. 3.

CHAPTER IV
FROM THE SCIENTIFIC REVOLUTION TO THE LIBERAL EXPECTATION

1. Maurice Cranston, *John Locke: A Biography* (London: Longmans, Green and Co. 1957), pp. 18–20.

2. Kenneth Dewhurst, *John Locke (1632–1704), Physician and Philosopher: A Medical Biography with an Edition of the Medical Notes in His Journals* (London: Wellcome Historical Medical Library, 1963), p. 34.

3. *Ibidem*, pp. 159–160.

4. Cranston, *John Locke*, pp. 39–40.

5. Kenneth Dewhurst, *Thomas Willis's Oxford Lectures* (Oxford: Sandford Publications, 1980), p. 44.

6. Cranston, *John Locke*, p. 25.

7. Dewhurst, *Thomas Willis's Oxford Lectures*, p. 44.

8. Kenneth Dewhurst, "An Oxford Medical Quartet: Sydenham, Willis, Locke, and Lower," in *Oxford Medicine: Essays on the Evolution of the Oxford Clinical School to Commemorate the Bicentary* [*sic*] *of the Radcliffe Infirmary 1770–1970,* Kenneth Dewhurst, ed. (Oxford: Sandford Publications, 1970), pp. 23–31 at p. 24.

9. Michael Hunter, "Alchemy, Magic and Moralism in the Thought of Robert Boyle," *British Journal of the History of Science* 23 (1990): 387–410; Michael Hunter, "Casuistry in Action: Robert Boyle's Confessional Interviews with Gilbert Burnet and Edward Stillingfleet, 1691," *Journal of Ecclesiastical History* 44, no. 1 (January 1993): 80–98.

10. Richard Lower, *DIATRIBAE THOMAE WILLISII Doct. Med. & Profess. Oxon. DE FEBRIBUS VINDICATIO ADVERSUS Edmundum De Meara Ormoniensem Hibernum M.D.* (London: Jo. Martyn & Ja. Allestry, 1665), published in facsimile as *Richard Lower's "Vindicatio": A Defence of the Experimental Method,* Kenneth Dewhurst, ed. (Oxford: Sandford Publications, 1983).

11. Dewhurst, *John Locke,* pp. 3–6.

12. Dewhurst, *Thomas Willis's Oxford Lectures,* p. 45.

13. Cranston, *John Locke,* pp. 59–64.

14. *Ibidem,* p. 41.

15. *Ibidem,* p. 74.

16. *Ibidem,* pp. 69–74.

17. *Ibidem,* pp. 74–87.

18. Richard Ashcraft, *Revolutionary Politics & Locke's Two Treatises of Government* (Princeton, N.J.: Princeton University Press, 1986), p. 22.

19. Dewhurst, *Thomas Willis's Oxford Lectures,* pp. 42–46.

20. Dewhurst, *John Locke,* p. 8.

21. Cranston, *John Locke,* p. 76.

22. Dewhurst, *Thomas Willis's Oxford Lectures,* pp. vii, ix, 42–46.

23. Webster, *The Great Instauration,* pp. 50–61.

24. *Ibidem,* p. 61.

25. *Ibidem,* pp. 166–169.

26. *Ibidem,* p. 88.

27. Cranston, *John Locke,* p. 116.

28. Michael Hunter [Birkbeck College, University of London], "Boyle *versus* the Galenists: An Aborted Revolution in Seventeenth-Century Medicine," unpublished paper read to the Triangle Workshop in the History of Science, Medicine and Technology, University of North Carolina at Chapel Hill, February 26, 1993.

29. George Wilson, "Robert Boyle," in George Wilson, *Religio Chemici. Essays,* Jessie A. Wilson, ed. (London: Macmillan, 1862), pp. 165–252 at pp. 232–233 (Microprint Landmarks of Science, viii p, [2] p l, 386 p, [2] p).

30. Lower, *Vindicatio,* pp. 197–202.

31. Dewhurst, *John Locke,* p. 34.

32. William Osler, "An Address on John Locke as a Physician," *Lancet,* no. 4025 (October 20, 1900): 1115–1123.

33. *A Bibliography of the Honourable Robert Boyle, Fellow of the Royal Society,* 2d ed., John F. Fulton, ed. (Oxford: Clarendon Press, 1961), pp. 99–100.

34. Cranston, *John Locke,* p. 354.

35. *Ibidem,* p. 361.
36. *A Bibliography of the Honourable Robert Boyle,* pp. 133–134.
37. Ashcraft, *Revolutionary Politics,* pp. 83–84.
38. Steven Shapin and Simon Schaffer, *Leviathan and the Air-Pump: Hobbes, Boyle, and the Experimental Life* (Princeton, N.J.: Princeton University Press, 1985), pp. 7–20.
39. Thomas Hobbes, *Dialogus Physicus de Natura Aeris . . . [A Physical Dialogue, or a Conjecture about the Nature of the Air taken up from Experiments recently made in London at Gresham College*], Simon Schaffer, trans., in Shapin and Schaffer, *Leviathon,* pp. 345–391.
40. Robert Boyle, *An Examen of the greatest part of Mr. Hobbes's Dialogus Physicus de Natura Aeris* in Robert Boyle, *The Works of the Honourable Robert Boyle in Six Volumes to which is prefixed the Life of the Author,* Thomas Birch, ed. (London: J. and F. Rivington *et alii,* 1744, 1772), presented in facsimile as Robert Boyle, *The Works,* Thomas Birch, ed. (Hildesheim, Germany: Georg Olms Verlagsbuchhandlung, 1965), pp. 189–242 at p. 190.
41. Robert Horwitz, "Introduction," in John Locke, *Questions concerning the Law of Nature,* Robert Horwitz and Jenny Strauss Clay, eds., Diskin Clay, trans. (Ithaca, N.Y.: Cornell University Press, 1990), pp. 1–10.
42. Hunter, "Boyle *versus* the Galenists."
43. Dewhurst, *John Locke,* p. 20.
44. Kenneth Dewhurst, *Dr. Thomas Sydenham (1624–1689): His Life and Original Writings* (Berkeley: University of California Press, 1966), pp. 33–34.
45. Dewhurst, *John Locke,* p. 20.
46. Cranston, *John Locke,* pp. 88–91.
47. Dewhurst, *Dr. Thomas Sydenham,* p. 60.
48. Thomas Sydenham, *Observationes Medicae,* presented as *The Works of Thomas Sydenham, M.D., Translated from the Latin Edition of Dr. Greenhill with a Life of the Author by R. G. Latham, M.D., etc. etc. etc.* (London: Sydenham Society, 1848), published in facsimile as Thomas Sydenham, *The Works of Thomas Sydenham* (Birmingham, Ala.: Classics of Medicine Library, 1979), pp. 27–28.
49. Dewhurst, *John Locke,* p. 41.
50. *Ibidem,* pp. 36–38.
51. Dewhurst, *Dr. Thomas Sydenham,* p. 39.
52. Dewhurst, *John Locke,* pp. 49–50.
53. Cranston, *John Locke,* pp. 160–168.
54. Sydenham, *Observationes Medicae,* p. 6.
55. Dewhurst, *John Locke,* pp. 57–58, 93–98.
56. *Ibidem,* pp. 57, 122–124.
57. Jerome J. Bylebyl, "The School of Padua: Humanistic Medicine in the Sixteenth Century," in *Health, Medicine, and Mortality in the Sixteenth Century,* Charles Webster, ed. (Cambridge, U.K.: Cambridge University Press, 1979), pp. 335–370.
58. Dewhurst, *John Locke,* pp. 51–54.
59. Dewhurst, *Dr. Thomas Sydenham,* p. 28.
60. Dewhurst, *John Locke,* pp. 295–311.
61. *Ibidem,* pp. 58–60.

62. *Ibidem*, p. 59.

63. Dewhurst, *Dr. Thomas Sydenham*, pp. 15–16.

64. *Ibidem*, p. 55.

65. *Ibidem*, p. 73.

66. Osler, "An Address on John Locke as a Physician."

67. Patrick Romanell, *John Locke and Medicine: A New Key Locke* (Buffalo, N.Y.: Prometheus Books, 1984), pp. 69–85.

68. Dewhurst, *John Locke*, pp. vii–viii.

69. Romanell, *John Locke*, pp. 160–162.

70. *Ibidem*, p. 160.

71. Dewhurst, *John Locke*, pp. vii–viii.

72. Romanell, *John Locke*, pp. 161–162.

73. *Ibidem*, pp. 29–31.

74. Peter Alexander, *Ideas, Qualities and Corpuscles: Locke and Boyle on the External World* (Cambridge, U.K.: Cambridge University Press, 1985), p. 7.

75. *Ibidem*, p. 73.

76. *Ibidem*, p. 82.

77. *Ibidem*, p. 23.

78. *Ibidem*, p. 57.

79. François Duchesneau, *L'Empirisme de Locke* [International Archives of the History of Ideas, vol. 57] (The Hague: Martinus Nijhoff, 1973), pp. 42–59.

80. Dewhurst, *Dr. Thomas Sydenham*, p. 73.

81. Thomas Sydenham, *De Arte Medica*, in Dewhurst, *Dr. Thomas Sydenham*, pp. 80–82.

82. John Locke, *An Essay concerning Humane Understanding* (London: Eliz. Holt for Thomas Basset, 1690), presented as John Locke, *An Essay concerning Human Understanding*, Peter H. Nidditch, ed. (Oxford: Clarendon Press, Oxford University Press, 1975), "The Epistle to the Reader," pp. 9–10.

83. *Ibidem*, "The Epistle to the Reader," p. 8.

84. *Ibidem*, book 1, chapter 1, para. 2, p. 44.

85. Romanell, *John Locke*, pp. 144–148.

86. Locke, *An Essay*, book 1, chapter 1, paras. 4, 6–7, pp. 44–47.

87. *Ibidem*, book 1, chapter 3, para. 3, p. 67.

88. *Ibidem*, book 1, chapter 2, para. 22, pp. 59–60.

89. *Ibidem*, book 1, chapter 2, para. 28, p. 65.

90. *Ibidem*, book 1, chapter 3, para. 2, p. 66.

91. *Ibidem*, book 1, chapter 3, para. 4, p. 68.

92. *Idem.*

93. *Idem.*

94. *Ibidem*, book 1, chapter 3, para. 12, p. 73.

95. *Ibidem*, book 1, chapter 3, para. 12, p. 74.

96. *Ibidem*, book 1, chapter 3, paras. 9–12, pp. 70–74.

97. *Ibidem*, book 1, chapter 3, paras. 22–26, pp. 81–83.

98. Dewhurst, *Thomas Willis's Oxford Lectures*, pp. 65–66.

99. *Ibidem*, p. 65.

100. Locke, *An Essay*, book 1, chapter 4, para. 8, pp. 87–88.

101. *Ibidem*, book 1, chapter 4, para. 12, p. 92.

102. *Ibidem*, book 1, chapter 4, para. 17, p. 95.

103. Ernst Mayr, "The Origins of Human Ethics" (1985), in *Toward a New Philosophy of Biology: Observations of an Evolutionist* (Cambridge, Mass.: Harvard University Press, 1988), pp. 75–91 at pp. 81–83.

104. Locke, *An Essay*, book 2, chapter 28, p. 353.

105. *Ibidem*, book 2, chapter 28, para. 12, pp. 356–357.

106. *Ibidem*, book 2, chapter 33, paras. 1–5, pp. 394–395.

107. *Ibidem*, book 2, chapter 33, paras. 6–17, pp. 396–400.

108. *Ibidem*, book 2, chapter 33, para. 18, p. 400.

109. *Idem.*

110. *Ibidem*, book 3, chapter 9, para. 16, pp. 484–485.

111. *Ibidem*, book 4, chapter 3, paras. 24–29, pp. 554–560.

112. *Ibidem*, book 4, chapter 3, para. 18, p. 549.

113. *Ibidem*, book 4, chapter 3, para. 18, pp. 549–550.

114. *Idem.*

115. *Ibidem*, book 4, chapter 3, para. 20, p. 552.

116. John Locke, *An Essay concerning Humane Understanding* (London: Eliz. Holt for Thomas Basset, 1690), presented as John Locke, *An Essay concerning Human Understanding*, Alexander Campbell Fraser, ed. (New York: Dover Publications [through special arrangement with Oxford University Press], 1959), vol. 2, p. 212, note 1.

117. John Locke, *An Essay*, Peter H. Nidditch, ed., book 4, chapter 12, para. 12, p. 647.

118. David Hartley, *Observations on Man, His Frame, His Duty, and His Expectations* (London: S. Richardson for James Leake and Wm. Frederick, 1749), presented in facsimile by Theodore L. Huguelet (Gainesville, Fla.: Scholar's Facsimiles & Reprints, 1966), part 1, p. 5.

119. Kathleen Wellman, *La Mettrie: Medicine, Philosophy, and Enlightenment* (Durham, N.C.: Duke University Press, 1992), pp. 149–161.

120. Elisha Bartlett, *An Essay on the Philosophy of Medical Science* (Philadelphia: Lea & Blanchard, 1844), pp. 10–25.

121. Worthington Hooker, *Physician and Patient; or, A Practical View of the Mutual Duties, Relations and Interests of the Medical Profession and the Community* (New York: Baker and Scribner, 1849), pp. 200–221.

122. Gilbert Blane, *Elements of Medical Logick, or Philosophical Principles of the Practice of Physick*, 3d ed. (London: Thomas and George Underwood, 1825), pp. 22, 73, 90, 219, 270–271.

123. *Ibidem*, p. 22.

124. *Ibidem*, pp. 270–271.

125. Richard Ashcraft, *Locke's Two Treatises of Government* (London: Unwin Hyman, 1987), p. 17.

126. Locke, *An Essay*, Alexander Campbell Fraser, ed., vol. 1, pp. 174–175.

127. *Ibidem*, vol. 1, pp. 174–176, at p. 175, note 2.

128. Hippocrates, *Ancient Medicine*, xx, pp. 53–55.

129. Dewhurst, *John Locke*, pp. 111, 112.

130. *Ibidem*, pp. 167, 176.

131. *Ibidem*, p. 193.

132. Samuel Johnson, page citation at "MANNA."

133. Dewhurst, *John Locke*, p. 224.

134. Cranston, *John Locke*, pp. 246–257.

135. Raymond Klibansky, "Preface," in John Locke, *Epistola de Tolerantia: A Letter on Toleration*, Raymond Klibansky, ed., and J. W. Gough, ed. and trans. (Oxford: Clarendon Press, 1968), p. xvi.

136. *Ibidem*, pp. xix–xx.

137. John Locke, *A Letter Concerning Toleration*, in Mario Montuori, *John Locke on Toleration and the Unity of God* (Amsterdam: J. C. Gieben, Publishers, 1983), p. 21.

138. *Ibidem*, pp. 97–99.

139. Peter Laslett, "Introduction," in John Locke, *Two Treatises of Government: A Critical Edition with an Introduction and Apparatus Criticus*, amended reprinting, Peter Laslett, ed. (New York: Mentor Books, by arrangement with Cambridge University Press, 1960 and 1963), pp. 58–64.

140. *Ibidem*, p. 75.

141. *Ibidem*, pp. 16–17.

142. John Locke, *Two Treatises of Government: In the Former, the False Principles and Foundation of Sir Robert Filmer, and His Followers, Are Detected and Overthrown. The Latter is an Essay Concerning the True Original, Extent, and End of Civil-Government* (London: Awnsham and John Churchill, 1698), presented as John Locke, *Two Treatises of Government: A Critical Edition with an Introduction and Apparatus Criticus*, amended reprinting, Peter Laslett, ed. (New York: Mentor Books, by arrangement with Cambridge University Press, 1960 and 1963), *Book I* or *First Treatise* . . . [F. T.], chapter 2, para. 7, p. 180.

143. *Ibidem, F. T.*, chapter 2, para. 6, p. 178.

144. *Ibidem, F. T.*, chapter 4, para. 42, pp. 205–206.

145. *Ibidem, F. T.*, chapter 5, para. 48, p. 210.

146. *Ibidem, F. T.*, chapter 6, para. p. 214.

147. *Idem.*

148. *Idem.*

149. *Ibidem, F. T.*, chapter 6, para. 55, p. 216.

150. *Ibidem, F. T.*, chapter 6, para. 56, pp. 217–218.

151. *Ibidem, F. T.*, chapter 6, para. 58, p. 219.

152. *Ibidem, F. T.*, chapter 9, para. 81, p. 240.

153. Laslett, "Introduction," p. 79.

154. John Locke, *Two Treatises, F. T.*, chapter 6, para. 54, pp. 215–216.

155. *Ibidem, F. T.*, chapter 9, para. 88, p. 244.

156. *Idem.*

157. *Ibidem, F. T.*, chapter 11, para. 147, p. 289.

158. *Ibidem, F. T.*, chapter 9, para. 88, p. 244.

159. *Ibidem, Two Treatises, Book II* or *Second Treatise* [*S. T.*], chapter 2, para. 6, p. 311.

160. *Ibidem, S. T.*, chapter 2, para. 12, p. 315.

161. *Ibidem, S. T.*, chapter 2, para. 14, p. 317.

162. *Ibidem, S. T.*, chapter 2, para. 14, pp. 317–318.

163. *Ibidem, S. T.*, chapter 3, paras. 19–21, pp. 321–323.

164. *Ibidem, S. T.*, chapter 7, paras. 90–92, pp. 369–371.

165. *Ibidem, S. T.*, chapter 5, para. 25, p. 327.

166. Geoffrey Keynes, *A Bibliography of Sir William Petty F.R.S. and of Observations on the Bills of Mortality by John Graunt F.R.S.* (Oxford: Clarendon Press, 1971), pp. 75–82.

167. Alessandro Roncaglia, *Petty: The Origins of Political Economy*, Isabella Cherubini, trans. (Armonk, N.Y.: M. E. Sharpe, 1985); Michael Perelman, *Classical Political Economy: Primitive Accumulation and the Social Division of Labor* (London: Rowman & Allanheld, 1983), pp. 67–71.

168. Peter Laslett, "John Locke and His Books," in Harrison and Laslett, *The Library of John Locke*, p. 25.

169. John Locke, *S. T.*, chapter 5, paras. 26–32, pp. 328–332.

170. *Ibidem, S. T.*, chapter 5, paras. 33–37, pp. 333–337.

171. *Ibidem, S. T.*, chapter 11, para. 135, p. 402; *Ibidem, S. T.*, chapter 15, para. 172, p. 429.

172. *Ibidem, S. T.*, chapter 6, para. 67, p. 355.

173. *Ibidem, S. T.*, chapter 6, paras. 79–80, pp. 362–363.

174. *Ibidem, S. T.*, chapter 9, para. 128, p. 397.

175. *Ibidem, S. T.*, chapter 16, para. 176, p. 433.

176. *Ibidem, S. T.*, chapter 16, para. 179, p. 435.

177. *Ibidem, S. T.*, chapter 16, paras. 177–196, pp. 433–444.

178. Cranston, *John Locke*, p. 155.

179. Locke, *Two Treatises, S. T.*, chapter 16, para. 189, pp. 440–441.

180. *Ibidem, S. T.*, chapter 18, para. 202, p. 449.

181. Laslett, "Introduction," p. 79.

182. Locke, *An Essay*, Peter H. Nidditch, ed., book 1, chapter 1, para. 6, p. 46.

183. Locke, *Questions*, pp. 119–137 at p. 119.

184. *Ibidem*, pp. 153–169 at p. 153.

185. Horwitz, "Introduction," in Locke, *Questions*, pp. 28–62.

186. Leo Strauss, *Natural Right and History* (Chicago: University of Chicago Press, 1953), pp. 165–166, 202–251; Richard H. Cox, *Locke on War and Peace* (Oxford: Oxford University Press, 1960), pp. xx, 139–147.

187. Paul E. Sigmund, *Natural Law in Political Thought* (Lanham, Md.: University Press of America, 1971), pp. 81–83.

188. Horwitz, "Introduction," pp. 6–11; Ruth W. Grant, *John Locke's Liberalism* (Chicago: University of Chicago Press, 1987), pp. 9–10.

189. Strauss, *Natural Right and History*, pp. 246–251.

190. Romanell, *John Locke*, pp. 29–34.

191. "The Celebrated Locke as a Physician," *Lancet*, no. 303 (June 20, 1829): 367.

192. Romanell, *John Locke*, pp. 32–33.

193. Mr. Cline, "Hunterian Oration," *Lancet* (no number) (February 22, 1824): 235–244 at p. 236.

194. George Wilkins, Surgeon, "To the Editor of *The Lancet*," *Lancet*, (no number) (March 14, 1824): pp. 358–360.

195. "A literary Gentleman of our acquaintance . . .," *Lancet*, no. 206 (August 11, 1827): 600.

196. John P----e, "Locke as a Physician," *Lancet*, no. 346 (April 17, 1830): 92–94 at p. 92.

197. *Ibidem*, pp. 92–94.

198. H. R. Fox Bourne, *The Life of John Locke* (New York: Harper & Brothers, 1876).

199. Romanell, *John Locke*, p. 39–40.

200. *Ibidem*, p. 83.

201. Cushing, *The Life of Sir William Osler*, pp. 672–680.

202. Osler, "An Address."

203. William Osler, "The Importance of Post-graduate Study," *Lancet*, no. 4011 (July 14, 1900): 73–75 at 74.

204. "Dr. Osler has found time . . .," *Lancet*, no. 4026 (October 27, 1900): 1216–1217.

CHAPTER V
FROM NONLIBERAL ALTERNATIVES TO THE LIBERAL REESTABLISHMENT

1. Romanell, *John Locke*, p. 96.

2. Chauncey D. Leake, "Introductory Essay on Percival's Medical Ethics," in Thomas Percival, *Percival's Medical Ethics* (Baltimore: Williams & Wilkins, 1927), pp. 1–2.

3. Thomas Percival, *Medical Ethics; or, a Code of Institutes and Precepts, Adapted to the Professional Conduct of Physicians and Surgeons* . . . (Manchester: S. Russell, 1803), presented as Thomas Percival, *Percival's Medical Ethics* (Baltimore: Williams & Wilkins, 1927), p. 147.

4. Code of Ethics of the American Medical Association (May 1847) (Philadelphia: T. K. and P. G. Collins, Printers, 1848), appended to Thomas Percival, *Percival's Medical Ethics*, pp. 218–238 at pp. 219–220.

5. Bernard, *An Introduction to the study of Experimental Medicine*, pp. 99–105 at p. 101.

6. Richard A. Meckel, *Save the Babies: American Public Health Reform and the Prevention of Infant Mortality, 1850–1829* (Baltimore: Johns Hopkins University Press, 1990), pp. 11–18.

7. John M. Eyler, "Mortality Statistics and Victorian Health Policy: Program and Criticism," *Bulletin of the History of Medicine* 50 (1976): 335–355 at 337.

8. Richard A. Meckel, *Save the Babies*, p. 18.

9. William Osler, "Chauvinism in Medicine," in William Osler, *Aequanimitas, with Other Addresses to Medical Students, Nurses and Practitioners of Medicine*, 3d ed. (New York: McGraw-Hill, no date), pp. 265–289.

10. *Ibidem*, pp. 265–271.

11. *Ibidem*, p. 268.

12. *Ibidem*, pp. 268–269.

13. *Ibidem*, p. 270.

14. *Ibidem*, pp. 271–274.

15. W. Bruce Fye, M.A., "William Osler's Departure from North America: The Price of Success," *New England Journal of Medicine* 320, no. 21 (May 25, 1989): 1425–1431.

16. Osler, "Chauvinism in Medicine," p. 288.

17. Osler, "Science and War," in Cushing, *The Life of Sir William Osler*, vol. 2, pp. 492–494.

18. Osler, "The Old Humanities and the New Science," p. 2.

19. *Idem.*

20. *Idem.*

21. Rudolf Virchow, *Cellular Pathology as Based upon Physiological and Pathological Histology*, trans. from 2d ed. of the original, Frank Chance, trans. (London: John Churchill, 1860), published in facsimile as Rudolph Virchow, *Cellular Pathology* (Birmingham, Ala.: Classics of Medicine Library, 1978), p. 27.

22. Ernst Mayr, *The Growth of Biological Thought: Diversity, Evolution, and Inheritance* (Cambridge, Mass.: Belknap Press of Harvard University Press, 1982), pp. 727–730.

23. Virchow, *Cellular Pathology*, pp. 13–14.

24. Paul Weindling, "Theories of the Cell State in Imperial Germany," in *Biology, Medicine and Society, 1840–1940*, Charles Webster, ed. (Cambridge, U.K.: Cambridge University Press, 1981), pp. 99–155 at p. 127.

25. Rudolf Virchow, "Atoms and Individuals" (1859), in *Disease, Life, and Man: Selected Essays by Rudolf Virchow*, Lelland J. Rather, ed. and trans. (Stanford, Calif.: Stanford University Press, 1958), pp. 120–141 at p. 130.

26. *Ibidem*, pp. 132–133.

27. Edward O. Wilson, *On Human Nature* (Cambridge, Mass.: Harvard University Press, 1978), pp. 76–77.

28. Jane Oppenheimer, "Driesch, Hans Adolf Eduard," in *Dictionary of Scientific Biography*, Charles Coulston Gillispie, ed. in chief (New York: Charles Scribner's Sons, 1971), vol. 4, pp. 186–189.

29. Natasha X. Jacobs, "From Unit to Unity: Protozoology, Cell Theory, and the New Concept of Life," *Journal of the History of Biology* 22, no. 2 (Summer 1989): 215–242 at 229, 235.

30. *Ibidem*, pp. 236–237.

31. Martin Heidenhain, *Plasma und Zelle. Erste Abteilung. Allgemeine Anatomie Der Lebendigen Masse* [*Plasma and Cell. First Section. General Anatomy of Living Matter*] (Jena: Gustav Fischer, 1907), p. 27 [R.H.S. translation.]

32. *Ibidem*, pp. 27–29.

33. *Ibidem*, pp. 29–110.

34. Hans Driesch, *The Science and Philosophy of the Organism: The Gifford Lectures Delivered before the University of Aberdeen in the Year 1907* [vol. 1] and . . . 1908 [vol. 2] (London: Adam and Charles Black, 1908), vol. 1, p. v.

35. *Ibidem*, vol. 1, p. 6.

36. *Ibidem*, vol. 1, pp. 322–323.

37. *Ibidem*, vol. 1, p. 324.

38. *Ibidem*, vol. 2, p. 44.

39. *Ibidem*, vol. 2, pp. 118–119.

40. *Ibidem*, vol. 2, p. 345.

41. *Ibidem*, vol. 1, p. vii.

42. Russell McCormmach, "On Academic Scientists in Wilhelmian Germany," in *Science and Its Public: The Changing Relationship*, Gerald Holton and William A. Blanpied, eds., Boston Studies in the Philosophy of Science, Robert S. Cohen

and Marx W. Wartofsky, eds., vol. 33 (Dordrecht, Holland: D. Reidel, 1976), pp. 157–171 at p. 158.

43. Eyler, "Mortality Statistics," p. 337.

44. Heinrich Schipperges, *Utopien der Medizin. Geschichte und Kritik der aertzlichen Ideologie des 19. Jahrhunderts* [*Utopias of Medicine: History and Critique of Medical Ideology of the Nineteenth Century*] (Salzburg: O. Mueller, 1968), pp. 42–44.

45. Weindling, "Theories of the Cell State in Imperial Germany," p. 155.

46. Virchow, "Atoms and Individuals," pp. 138–139.

47. Virchow, *Cellular Pathology*, p. x.

48. Lelland J. Rather, "Harvey, Virchow, Bernard, and the Methodology of Science," in Virchow, *Disease, Life, and Man*, pp. 1–25 at p. 21.

49. Rudolf Virchow, "Scientific Method and Therapeutic Standpoints," (1849) in *ibidem*, pp. 40–66 at p. 44.

50. Oskar Hertwig, *Die Lehre vom Organismus und ihre Beziehung zur Sozialwissenschaft* [*The Doctrine of the Organism and Its Relationship to Social Science*] (Jena: Gustav Fischer, 1899).

51. Oskar Hertwig, *Der Staat als Organismus, Gedanken zur Entwicklung der Menschheit* [*The State as Organism: Thoughts on the Development of Mankind*] (Jena, 1922).

52. Weindling, "Theories of the Cell State in Imperial Germany," p. 119.

53. *Ibidem*, p. 105.

54. *Idem.*

55. Charles Darwin, *The Descent of Man, and Selection in Relation to Sex*, 2d ed. (London: John Murray, 1877), vol. 21 in *The Works of Charles Darwin*, Paul H. Barrett and R. B. Freeman, eds. (London: William Pickering, 1989), part 1, pp. 138–148.

56. Degler, *In Search of Human Nature*, pp. 3–17 at p. 11.

57. Michael Ruse, *Taking Darwin Seriously: A Naturalistic Approach to Philosophy* (Oxford: Basil Blackwell, 1986), p. 44.

58. Herbert Spencer, "The Social Organism," in *Essays: Scientific, Political, and Speculative* (London: Williams and Norgate, 1868), pp. 384–428.

59. *Ibidem*, pp. 388–391.

60. *Ibidem*, pp. 398–399.

61. *Ibidem*, pp. 423–428.

62. Daniel Gasman, *The Scientific Origins of National Socialism: Social Darwinism in Ernst Haeckel and the German Monist League* (London: Macdonald and Company; and New York: American Elsevier, 1971), pp. ix–30, 82–105, 147–182; Paul Weindling, *Health, Race and German Politics between National Unification and Nazism, 1870–1945* (Cambridge, U.K.: Cambridge University Press, 1989), pp. 40–48; Ernst Haeckel, *Die Welträtsel: Gemeinverständliche Studien über monistische Philosophie* [*The Riddles of the Universe: Popular Essays on Monistic Philosophy*] (Berlin: Akademie-Verlag, 1961); *Haeckel, The Riddle of the Universe at the Close of the Nineteenth Century*, Joseph McCabe, trans. (New York and London: Harper and Brothers, 1902).

63. Ruse, *Taking Darwin Seriously*, pp. 82–86.

64. David D. Anderson, *William Jennings Bryan* (Boston: Twayne Publishers, 1981), pp. 178–179.

65. *Ibidem*, p. 131.

66. LeRoy Ashby, *William Jennings Bryan: Champion of Democracy* (Boston: Twayne Publishers, 1987), pp. 183–184.

67. Anderson, *William Jennings Bryan*, p. 132.

68. Ashby, *William Jennings Bryan*, 159–160.

69. *Ibidem*, pp. 183–184.

70. Auckland Geddes, "Social Reconstruction and the Medical Profession," in *Contributions to Medical and Biological Research Dedicated to Sir William Osler, Bart., M.D., F.R.S., in Honour of His Seventieth Birthday, July 12, 1919, by His Pupils and Co-workers* (New York: Paul B. Hoeber, 1919), vol. 1, pp. 70–79.

71. *Ibidem*, pp. 70–71.

72. *Ibidem*, pp. 71–72.

73. Aristotle, *Politics*, book 1, chapter 1, para. 2, p. 5.

74. Plato, *The Republic*, Paul Shorey, trans. (Cambridge, Mass.: Harvard University Press, 1978), book 5, chapters 8–9, pp. 459–469.

75. Aristotle, *Politics*, book 7, chapter 14, pp. 617–625.

76. Herbert Spencer, *The Principles of Biology* (London: Williams and Norgate, 1864 [vol. 1] and 1867 [vol. 2]), vol. 2, pp. 495–497.

77. Darwin, *Descent of Man*, p. 139.

78. *Ibidem*, pp. 138–139.

79. Leonard Darwin, *The Need for Eugenic Reform* (New York: D. Appleton, 1926), published in facsimile as Leonard Darwin, *The Need for Eugenic Reform*, vol. 4 in The History of Hereditarian Thought Series, Charles Rosenberg, ed. (New York: Garland Publishing, 1984), p. 130.

80. *Ibidem*, p. 171.

81. *Ibidem*, pp. 171–172.

82. Daniel J. Kevles, *In the Name of Eugenics: Genetics and the Uses of Human Heredity* (Berkeley: University of California Press, 1985), pp. 110–112, 329–330.

83. "Buck v. Bell, Superintendent. Error to the Supreme Court of Appeals of the State of Virginia. no. 292. Argued April 22, 1927.— Decided May 2, 1927," in *United States Reports*, vol. 274, *Cases Adjudged in the Supreme Court at October Term, 1926, Ernest Knaebel, Reporter* (Washington, D.C.: U.S. Government Printing Office, 1928), pp. 200–208.

84. "To Halt the Imbecile's Perilous Line," *The Literary Digest* 93, no. 8 (May 21, 1927): 11.

85. "Buck v. Bell, Superintendent."

86. *Idem.*

87. Benno Mueller-Hill, *Murderous Science: Elimination by Scientific Selection of Jews, Gypsies, and Others, Germany, 1933–1945*, George R. Fraser, trans. (Oxford: Oxford University Press, 1988), p. 8.

88. Robert Proctor, *Racial Hygiene: Medicine under the Nazis* (Cambridge, Mass.: Harvard University Press, 1988), p. 18.

89. J.B.S. Haldane, *Heredity and Politics* (New York: Norton, 1938), p. 167.

90. *Trials of War Criminals before the Nuernberg Military Tribunals under Control Council Law No. 10* (Nuernberg: Nuernberg Military Tribunals, October 1946, to April 1949), vol. 1: "The Medical Case," pp. 15–17.

91. *Ibidem*, vol. 1: "The Medical Case," pp. 11–15.

92. *Ibidem,* vol. 2: "The Medical Case"; "The Milch Case," pp. 138–140.

93. Robert Jay Lifton, *The Nazi Doctors: Medical Killing and the Psychology of Genocide* (New York: Basic Books, 1986), p. 430.

94. *Ibidem,* pp. 437–438.

95. *Ibidem,* p. 488.

96. *Trials of War Criminals,* vol. 2: "The Medical Case"; "The Milch Case," pp. 181–184.

97. "Declaration of Geneva, World Medical Association, 1948," in *Encyclopedia of Bioethics,* Warren T. Reich, ed. in chief (New York: Free Press, 1978), vol. 4, p. 1749.

98. David Joravsky, *Soviet Marxism and Natural Science: 1917–1932* (New York: Columbia University Press, 1961), pp. 196–310; Zhores A. Medvedev, *The Rise and Fall of T. D. Lysenko,* I. Michael Lerner, trans., Lucy G. Lawrence, ed. asst. (New York: Columbia University Press, 1969); David Joravsky, *The Lysenko Affair* (Cambridge, Mass.: Harvard University Press, 1970); William Broad and Nicholas Wade, *Betrayers of the Truth* (New York: Simon and Schuster, 1982), p. 189; Valery N. Soyfer, "New Light on the Lysenko Era," *Nature* 339, (June 8, 1989): 415–420 (with errata, vol. 340, (August 31, 1989): 732).

99. Lifton, *The Nazi Doctors,* p. 9.

CHAPTER VI
FROM ALTRUISM TO ACTIVISM

1. Albert Schweitzer, *Out of My Life and Thought: An Autobiography,* Antje Bultmann Lemke, trans. (New York: Henry Holt, 1933, 1949, 1990), pp. 81–82.

2. *Ibidem,* p. 115.

3. *Ibidem,* p. 82.

4. *Ibidem,* p. 34; Albert Schweitzer, "Author's Preface to the First English Edition," J. P. Naish, trans., in *The Philosophy of Civilization* (1923, 1932), C. T. Campion, trans. (Buffalo, N.Y.: Prometheus Books, 1987), p. xi.

5. Schweitzer, *Out of My Life and Thought,* pp. 147–148.

6. *Ibidem,* pp. 154–155.

7. Albert Schweitzer, *The Rights of the Unborn and the Peril Today* (Chicago: Albert Schweitzer Education Foundation, 1958); idem, *On Nuclear War and Peace,* Homer A. Jack, ed. (Elgin, Ill.: Brethren Press, 1988).

8. Albert Schweitzer, "Author's Preface to the First English Edition," J. P. Naish, trans., in *The Philosophy of Civilization,* pp. xi–xii.

9. Schweitzer, *The Philosophy of Civilization,* p. 144.

10. *Ibidem,* pp. 143–189.

11. *Ibidem,* pp. 198–207.

12. *Ibidem,* pp. 213–220.

13. *Ibidem,* pp. 235–248.

14. *Ibidem,* p. 4.

15. *Ibidem,* pp. 221–234.

16. *Ibidem,* p. 4.

17. *Ibidem,* pp. 151–162.

18. Immanuel Kant, "Metaphysical Foundations of Morals" (1785), in *The Phi-*

losophy of Kant: Immanuel Kant's Moral and Political Writings, Carl J. Friedrich, ed. (New York: Modern Library, 1949, 1977), pp. 140–208 at p. 170.

19. Schweitzer, *The Philosophy of Civilization*, pp. 181–185.

20. *Ibidem*, pp. 221–234.

21. Wilhelm Stern, *Kritische Grundlegung der Ethik als positiver Wissenschaft* [*Critical Foundation of Ethics as a Positive Science*] (Berlin: Ferd. Dümmlers Verlagsbuchhandlung, 1897).

22. Wilhelm Stern, *Die allgemeinen Principien der Ethik auf naturwissenschaftlicher Basis: Vortrag, gehalten in der Naturwissenschaftlichen Abtheilung der Berliner Finkenschaft am 4. Dezember 1900* [*The General Principles of Ethics on a Natural-Science Basis: A Lecture Delivered in the Natural Science Section of the Berlin Finkenschaft on the Fourth of December 1900*] (Berlin: Ferd. Dümmlers Verlagsbuchhandlung, 1901).

23. Schweitzer, *The Philosophy of Civilization*, pp. 259–260.

24. *Ibidem*, p. 76.

25. *Ibidem*, pp. 291–292.

26. *Ibidem*, pp. 292–293.

27. *Ibidem*, p. 293.

28. *Ibidem*, p. 309.

29. *Ibidem*, p. 318.

30. John Stuart Mill, "Utilitarianism" (1861, 1863), in John Stuart Mill and Jeremy Bentham, *Utilitarianism and Other Essays*, Alan Ryan, ed. (Harmondsworth, U.K.: Penguin Books, 1987), pp. 272–338 at p. 324.

31. John R. Baker, *Science and the Planned State* (London: George Allen & Unwin, 1945), pp. 105–108.

32. Schweitzer, *The Philosophy of Civilization*, pp. 310–311.

33. *Ibidem*, p. 311.

34. *Ibidem*, pp. 317–318.

35. *Ibidem*, pp. 320–322.

36. Joel Mattison, "Lessons from Lambaréné: Part II," *Bulletin of the American College of Surgeons* 77, no. 10 (October 1992): 7–15.

37. Schweitzer, *The Philosophy of Civilization*, pp. 324–325.

38. *Ibidem*, pp. 326–328.

39. *Ibidem*, pp. 330–338.

40. *Ibidem*, pp. 342–344.

41. *Ibidem*, pp. 29, 332–333, 341–342.

42. *Ibidem*, p. 327.

CHAPTER VII
LIFE-SCIENCES LIBERALISM IN ABSTRACT AND COMPETITION

1. Thucydides, *History of the Peloponnesian War*, Charles Forster Smith, trans. (Cambridge, Mass.: Harvard University Press, 1980).

2. Saint Augustine, *The City of God Against the Pagans*, George E. McCracken (vol. 1), William M. Green (vol. 2), William Chase Greene (vol. 6), trans. (London: William Heinemann; Cambridge, Mass.: Harvard University Press, 1957, 1963, 1969).

3. Thomas Aquinas, *On Princely Government: To the King of Cyprus* in *Selected*

Political Writings, A. P. d'Entreves, ed., J. G. Dawson, trans. (Oxford: Basil Blackwell, 1959).

4. Marsilius of Padua, *Defensor Pacis* in *Marsilius of Padua, The Defender of Peace, Volume II: The Defensor Pacis*, Alan Gewirth, ed. and trans. (New York: Columbia University Press, 1956).

5. Niccolò Machiavelli, *The Prince*, Christian Gauss, ed., Luigi Ricci and E.R.P. Vincent, trans. (New York: Mentor Books, New American Library of World Literature, 1952).

6. Thomas Hobbes, *Leviathan, or The Matter, Forme, & Power of a Common-wealth Ecclesiasticall and Civill* (London: Andrew Crooke, 1651), presented as Thomas Hobbes, *Leviathan*, C. B. Macpherson, ed. (London: Penguin Books, Ltd., 1968).

7. Jean-Jacques Rousseau, *On the Social Contract* (1762), in *On the Social Contract with Geneva Manuscript and Political Economy*, Roger D. Masters, ed., Judith R. Masters, trans. (New York: St. Martin's Press, 1978).

8. Georg Wilhelm Friedrich Hegel, *The Philosophy of History*, rev. ed. with prefaces by Charles Hegel and J. Sibree, trans. (New York: Willey Book Co., 1944; Georg Wilhelm Friedrich Hegel, *Hegel's Philosophy of Right*, T. M. Knox, ed. and trans. (Oxford: Clarendon Press, 1967).

9. Carl von Clausewitz, *On War* [*Vom Kriege* (1832)], Anatol Rapoport, ed., J. J. Graham, trans. (Harmondsworth, U.K.: Penguin Books, 1968).

10. Max Weber, *From Max Weber: Essays in Sociology*, H. H. Gerth and C. Wright Mills, eds. and trans. (New York: Oxford University Press, 1946).

11. Reinhold Niebuhr, *Moral Man and Immoral Society: A Study in Ethics and Politics* (New York: Charles Scribner's Sons, 1932).

12. Edward Hallett Carr, *The Twenty Year's Crisis, 1919–1939: An Introduction to the Study of International Relations* (1939, 1945), 2d ed. (New York: Harper Torchbooks, Harper & Row, Publishers, 1964).

13. George F. Kennan, *American Diplomacy*, expanded edition (Chicago: University of Chicago Press, 1984); idem, "Morality and Foreign Policy," *Foreign Affairs* 64, no. 2 (Winter 1985/86): 205–218.

14. Hans J. Morgenthau, *Scientific Man vs. Power Politics* (Chicago: Phoenix Books, University of Chicago Press, 1946); idem, *Politics among Nations: The Struggle for Power and Peace* (New York: Alfred A. Knopf, 1948); idem, *In Defense of the National Interest: A Critical Examination of American Foreign Policy* (Washington, D.C.: University Press of America, 1982).

15. Thomas Hobbes, *Leviathan*, part 2, chapter 29, p. 375.

16. Milton Friedman, "[A Friedman Doctrine—]The Social Responsibility of Business Is to Increase Its Profits," *New York Times Magazine*, September 13, 1970, pp. 32ff.

17. Robert Hunt Sprinkle, "States, Corporations, and the Ethics of Political Realism," in "Political Ethics and the Life Sciences" (Ph.D. diss., Princeton University, 1990), pp. 181–317.

18. Jeremy Bentham, *A Fragment on Government and an Introduction to the Principles of Morals and Legislation* (1776, 1789), Wilfrid Harrison, ed. (Oxford: Basil Blackwell, 1948).

19. James Mill, *Elements of Political Economy*, 2d ed., rev. and corr. (London: Printed for Baldwin, Craddock, and Joy, 1824).

20. John Stuart Mill, *On Liberty* (1859), Currin V. Shields, ed. (New York: Macmillan, 1956).

21. Mill, *Utilitarianism*, p. 326.

CHAPTER VIII
PROTECTING THE STATE

1. Walter B. Beals, *The First German War Crimes Trial: Chief Judge Walter B. Beals' Desk Notebook of the Doctors' Trial, Held in Nuernberg, Germany, December, 1945 to August, 1947*, 2d ed., W. Paul Burman, ed. (Chapel Hill, N.C.: Documentary Publications, 1985), pp. 121–198.

2. *Ibidem*, p. ii.

3. "Head of E.P.A. Bars Nazi Data in Study on Gas," *New York Times*, March 23, 1988, p. A1; "EPA Bars Use of Nazis' Human Test Data after Scientists Object," *Washington Post*, March 24, 1988, p. A17.

4. "Minnesota Scientist Plans to Publish Nazi Experiment on Freezing," *New York Times*, May 12, 1988, p. A28.

5. *Ibidem*; "Nazi Scientists and Ethics of Today," *New York Times*, May 21, 1989, p. A34.

6. John W. Powell, "Japan's Biological Weapons: 1930–1945. A Hidden Chapter in History," *Bulletin of the Atomic Scientists* 37 (October 1981): 44–52; Sheldon Harris, "Japanese Biological Warfare Experiments and Other Atrocities in Manchuria, 1932–1945, and the Subsequent United States Cover Up: A Preliminary Assessment," *Crime, Law and Social Change* 15 (1991): 171–199.

7. Powell, "Japan's Biological Weapons."

8. Harris, "Japanese Biological Warfare," p. 191.

9. Harris, "Japanese Biological Warfare."

10. Arnold C. Brackman, *The Other Nuremberg: The Untold Story of the Tokyo War Crimes Trials* (New York: William Morrow, 1987), pp. 197–198.

11. Harris, "Japanese Biological Warfare," p. 190.

12. Harris, "Japanese Biological Warfare."

13. Robert Harris and Jeremy Paxman, *A Higher Form of Killing: The Secret Story of Chemical and Biological Warfare* (New York: Hill and Wang, 1982), pp. 86–88.

14. Powell, "Japan's Biological Weapons."

15. *Ibidem.*

16. Harris, "Japanese Biological Warfare," pp. 186–188.

17. Charles Piller and Keith R. Yamamoto, *Gene Wars: Military Control over the New Genetic Technologies* (New York: Beech Tree Books, William Morrow, 1988), p. 41.

18. Harris, "Japanese Biological Warfare," p. 195, note 11.

19. Morimura Seiichi, [*A Kuma no Moshaku, Dai 731 Butai* or *Akuma no Hoshouku* or *The Devil's Gluttony*] *Section 731 en Mandchourie: Experimentations japonaises sur des cobayes humains pendant la Seconde Guerre mondiale*, Paul Couturiau, trans. and adapter from English (Monaco: Editions du Rocher, 1981, 1985).

20. Harris, "Japanese Biological Warfare," p. 192.

21. Powell, "Japan's Biological Weapons."

22. National Commission for the Protection of Human Subjects of Biomedical and Behavioral Research, *Research Involving Prisoners* (Bethesda, Md.: Department of Health, Education, and Welfare, 1976), (OS)76–131 and (OS)76–132.

23. National Commission for the Protection of Human Subjects of Biomedical and Behavioral Research, *Research Involving Those Institutionalized as Mentally Infirm* (Bethesda, Md.: Department of Health, Education, and Welfare, 1978), (OS)78–0006 and (OS)78–0007.

24. James H. Jones, *Bad Blood: The Tuskegee Syphilis Experiment* (New York: Free Press, 1981).

25. Henry K. Beecher, "Ethics and Clinical Research," *New England Journal of Medicine* 274, no. 24 (June 16, 1966): 1354–1360.

26. David J. Rothman, "Ethics and Human Experimentation: Henry Beecher Revisited," *New England Journal of Medicine* 317, no. 19 (November 5, 1987): 1195–1199.

27. Victor L. Yu, "*Serratia Marcescens*: Historical Perspective and Clinical Review," *New England Journal of Medicine* 300, no. 16 (April 19, 1979): 887–891.

28. *Biological Testing Involving Human Subjects by the Department of Defense, 1977: Hearings before the Subcommittee on Health and Scientific Research of the Committee on Human Resources, United States Senate, Ninety-fifth Congress, First Session on Examination of Serious Deficiencies in the Defense Department's Efforts to Protect the Human Subjects,* [*sic*] *of Drug Research, March 8 and May 23, 1977* (Washington, D.C.: U.S. Government Printing Office, 1977), p. 125.

29. Leonard A. Cole, *Clouds of Secrecy: The Army's Germ Warfare Tests over Populated Areas* (Totowa, N.J.: Rowman & Littlefield, 1988), pp. 78–79.

30. Richard P. Wheat, Anne Zuckerman, and Lowell Rantz, "Infection due to Chromobacteria: Report of Eleven Cases," *Archives of Internal Medicine* 88, no. 6 (October 1951): 461–466.

31. Burton A. Waisbren, "Bacteremia Due to Gram-negative Bacilli Other Than the Salmonella," *Archives of Internal Medicine,* 88, no. 6 (October 1951): 467–488.

32. Cole, *Clouds of Secrecy,* pp. 82–83.

33. *Biological Testing,* pp. 143–146.

34. Cole, *Clouds of Secrecy,* pp. 82–83.

35. *Biological Testing,* p. 15.

36. *Ibidem,* pp. 125–131.

37. Cole, *Clouds of Secrecy,* p. 84, note 1.

38. *Biological Testing.*

39. Cole, *Clouds of Secrecy,* pp. 75–104.

40. "Nuremberg Code, 1946 [*sic*]," in *Encyclopedia of Bioethics,* Warren T. Reich, ed. in chief (New York: Free Press, 1978), vol. 4, pp. 1764–1765.

41. Cole, *Clouds of Secrecy,* p. 33.

42. *Biomedical and Behavioral Research, 1975: Joint Hearings before the Subcommittee on Health of the Committee on Labor and Public Welfare and the Subcommittee on Administrative Practice and Procedure of the Committee on the Judiciary, United States Senate, Ninety-fourth Congress, First Session on Human-use Experimentation Programs of the Department of Defense and Central Intelligence Agency and S. 2515 To Amend the Public Health Service Act To Establish the President's Commission for the Protection of Human*

Subjects Involved in Biomedical and Behavioral Research, and for Other Purposes, September 10, 12; and November 7, 1975 (Washington, D.C.: U.S. Government Printing Office, 1976), p. 297.

43. *Ibidem*, pp. 879–905.

44. *Ibidem*, pp. 970–972.

45. *Ibidem*, p. 971.

46. *Ibidem*, p. 972.

47. *Ibidem*, p. 911.

48. *Ibidem*, pp. 985–990.

49. *Ibidem*, pp. 992–1132.

50. Jessica Mathews, "Secrecy's Radioactive Legacy," *New York Times*, January 5, 1994, p. A19; Tim Weiner, "C.I.A. Seeks Documents from Its Radiation Tests," *New York Times*, January 5, 1994, p. A11; "Nuclear Guinea Pigs," *New York Times*, January 5, 1994, p. A14.

51. Faina A. Abramova, Lev M. Grinberg, Olga V. Yampolskaya, and David H. Walker (communicated by Mathew Meselson, December 14, 1992), "Pathology of Inhalational Anthrax in 42 Cases from the Sverdlovsk Outbreak of 1979," *Proceedings of the National Academy of Sciences USA* 90 (March 15, 1993): 2291–2294.

52. Raymond L. Garthoff, *Détente and Confrontation: American-Soviet Relations from Nixon to Reagan* (Washington, D.C.: Brookings Institution, 1985).

53. Robert H. Sprinkle, "Bioweaponry and 'Life Sciences Liberalism,'" *Annals of the New York Academy of Sciences* 666 (December 31, 1992): 88–99.

CHAPTER IX
PURSUING THE NATIONAL POLITICAL ADVANTAGE

1. Charter of the United Nations, Article 2, Paragraph 1, in *Basic Documents in International Law and World Order*, 2d ed., Burns H. Weston, Richard A. Falk, and Anthony A. D'Amato, eds. (St. Paul, Minn.: West Publishing Co., 1990), pp. 16–32 at p. 16.

2. Charter of the United Nations, at p. 10.

3. Richard W. Nelson, "Current Developments," *The American Journal of International Law* 80, no. 4 (October 1986): 973–983.

4. Robert C. Johansen, "The Reagan Administration and the U.N.: The Costs of Unilateralism," *World Policy Journal* 3 no. 4 (Fall 1986): 601–641.

5. Gertrude Samuels, "The 20 Per Cent Solution: Kassebaum vs. Moynihan on the UN," *The New Leader*, May 5–19, 1986, pp. 7–10.

6. Elisabeth Zoller, "The 'Corporate Will' of the United Nations and the Rights of the Minority," *American Journal of International Law* 81, no. 3 (July 1987): 610–634 at 610–614.

7. *Ibidem* at pp. 633–634.

8. Chris Adams, "U.S. Lags on Payments to World Groups: Policy Disputes, Budget Deficit Blamed; Agencies Respond by Reducing Services," *Washington Post*, August 17, 1988, p. A16.

9. Don Shannon, "Reagan Releases U.N. Payments: Cites Fiscal Reforms, Peace Role; $188 Million Promised by Oct. 1," *Los Angeles Times*, September 14, 1988, p. 1.

10. Robert Pear, "Baker Would Ask Cutoff of Funds if U.N. Agencies Upgrade P.L.O.," *New York Times*, May 2, 1989, p. A12.

11. *U.S. Contributions to International Organizations: Hearing before the Subcommittees on Human Rights and International Organizations and on International Operations of the Committee on Foreign Affairs, House of Representatives, One Hundredth Congress, Second Session, February 23, 1988* (Washington, D.C.: U.S. Government Printing Office, 1988), pp. 31–32.

12. Richard Bernstein, "Withheld Dues Said to Cause W.H.O. Crisis," *New York Times*, December 18, 1986, p. A17.

13. Marjory Dam, "What the United States Owes WHO," *Washington Post*, October 22, 1987, p. A22.

14. Barry R. Bloom, "A New Threat to World Health," *Science* 239, no. 4835 (January 1, 1988): 9.

15. Barry R. Bloom, pers. comm., Princeton, N.J., March 9, 1989.

16. *United States Participation in the UN: Report by the President to the Congress for the Year 1986*, Department of State Publication 9557 (Washington, D.C.: United States Department of State, 1987), pp. 217–218.

17. *Ibidem*, pp. 227–228.

18. *Ibidem*, pp. 226–227.

19. *Ibidem*, p. 228.

20. *Ibidem*, pp. 221–222.

21. *Ibidem*, p. 222.

22. *Ibidem*, p. 219.

23. *Ibidem*, pp. 223–225.

24. *Ibidem*, p. 224.

25. *United States Participation in the UN: Report by the President to the Congress for the Year 1987*, Department of State Publication 9671 (Washington, D.C.: United States Department of State, 1988), p. 199.

26. *Ibidem*, p. 203.

27. *Ibidem*, pp. 203–204.

28. *Ibidem*, p. 204.

29. *Ibidem*, p. 202.

30. *Ibidem*, p. 204.

31. *Idem.*

32. *Idem.*

33. *Ibidem*, p. 278.

CHAPTER X
AGGRANDIZING THE CORPORATION

1. Hobbes, *Leviathan*, part 4, chapter 47, p. 714.

2. Adam Smith, *An Inquiry into the Nature and Causes of the Wealth of Nations*, 4th ed., R. H. Campbell, A. S. Skinner, W. B. Todd, eds. (Oxford: Clarendon Press, 1976), vol. 1, book 1, chapter 1, pp. 26–27.

3. Adam Smith, *The Theory of Moral Sentiments*, D. D. Raphael, A. L. Macfie, eds. (Oxford: Clarendon Press, 1976), part 3, chapter 3, paras. 5–6, pp. 137–138.

4. *Ibidem*, part 3, chapter 4, para. 12, p. 160.

5. *Ibidem*, part 7, section 3, chapter 2, para. 2, p. 318.

6. Milton Friedman with the assistance of Rose D. Friedman, *Capitalism and Freedom* (Chicago: University of Chicago Press, 1962), p. 133.

7. Milton Friedman, "[A Friedman Doctrine—]The Social Responsibility of Business Is to Increase Its Profits."

8. Thomas Donaldson, *Corporations and Morality* (Englewood Cliffs, N.J.: Prentice Hall, 1982), pp. 18–34.

9. *H.R. 4502 and S. 1966, The Biotechnology Competitiveness Act*, p. 208.

10. Randall S. Bock, "The Pressure to Keep Prices High at a Walk-in Clinic: A Personal Experience," *New England Journal of Medicine* 319, no. 12 (September 22, 1988): 785–787.

11. David U. Himmelstein and Steffie Woolhandler, "The Corporate Compromise: A Marxist View of Health Maintenance Organizations and Prospective Payment," *Annals of Internal Medicine* 109 (September 15, 1988): 494–501.

12. Mike Muller, *The Health of Nations: A North-South Investigation* (London: Faber and Faber, 1982), pp. 26–29; Milton Silverman, Mia Lydecker, and Philip R. Lee, *Bad Medicine: The Prescription Drug Industry in the Third World* (Stanford, Calif.: Stanford University Press, 1992), pp. 86–106.

13. Mike Muller, *The Health of Nations*, pp. 30–33; Silverman, Lydecker, and Lee, *Bad Medicine*, p. 27.

14. Mike Muller, *The Health of Nations*, pp. 34–35.

15. *Ibidem*, p. 35.

16. Silverman, Lydecker, and Lee, *Bad Medicine*, pp. 39–40, 313.

17. Manuel G. Velasquez, *Business Ethics: Concepts and Cases* (Englewood Cliffs, N.J.: Prentice Hall, 1982), pp. 37–42.

18. Barry I. Castleman, "The Double Standard in Industrial Hazards," in *The Export of Hazard: Transnational Corporations and Environmental Control Issues*, Jane H. Ives, ed. (Boston: Routledge & Kegan Paul, 1985), pp. 60–89 at p. 65.

19. Brooke T. Mossman and J. Bernard L. Gee, "Asbestos-Related Diseases," *New England Journal of Medicine* 320, no. 26 (June 29, 1989): 1721–1730.

20. Velasquez, *Business Ethics*, pp. 94–96; Richard T. DeGeorge, "Ethical Responsibilities of Engineers in Large Organizations: The Pinto Case," *Business and Professional Ethics Journal* 1, no. 1 (1981): 1–14.

21. George Pearsall, Sc. D., Duke University, pers. comm., January 27, 1993.

22. Kenneth E. Goodpaster, "The Concept of Corporate Responsibility," in *Just Business: New Introductory Essays in Business Ethics*, Tom Regan, ed. (Philadelphia: Temple University Press, 1983), pp. 292–323 at pp. 292–294.

23. Barry Meier, "Data Show G. M. Knew for Years of Risk in Pickup Trucks' Design: Company Says Changes Were Not Tied to Safety," *New York Times*, November 17, 1992, pp. A1, A10.

24. W. Eugene Smith and Aileen M. Smith, *Minamata*, containing Harada Masazumi with Aileen M. Smith, "Minamata Disease: A Medical Report" (New York: Holt, Rinehart and Winston, 1975), pp. 26–47.

25. *Ibidem*, pp. 178–191.

26. *Ibidem*, pp. 122–123.

27. *Ibidem*, p. 33.

28. Valerie A. Fildes, *Breasts, Bottles, and Babies: A History of Infant Feeding* (Edinburgh, U.K.: Edinburgh University Press, 1986), pp. 268–277.

29. *Ibidem*, p. 399.

30. Bertrand Russell, *A History of Western Philosophy* (New York: Touchstone Books, Simon and Schuster, 1945, 1972), pp. 686–687.

31. Fildes, *Breasts, Bottles,* p. 276.

32. *Ibidem,* pp. 274–75.

33. Andrew Chetley, *The Politics of Baby Foods: Successful Challenges to an International Marketing Strategy* (London: Frances Pinter, 1986), pp. 9–22.

34. *Contemporary Patterns of Breast-feeding: Report on the WHO Collaborative Study on Breast-feeding* (Geneva: World Health Organization, 1981), pp. 4–6.

35. Jean-Pierre Habicht, Julie DaVanzo, and William P. Butz, "Mother's Milk and Sewage: Their Interactive Effects on Infant Mortality," *Pediatrics* 81, no. 3 (March 1988): 456–461; Allan S. Cunningham, Derrick B. Jelliffe, and E. F. Patrice Jelliffe, "Breast-feeding and Health in the 1980s: A Global Epidemiological Review," *Journal of Pediatrics* 118, no. 5 (May 1991): 659–666.

36. Chetley, *Politics of Baby Food,* p. 3.

37. Anthony Sampson, *The Sovereign State of ITT* (New York: Stein and Day, 1973).

38. Christopher Tugendhat, *The Multinationals,* 2d ed. (London: Penguin, 1973), p. 22, as cited in Chetley, *Politics of Baby Food,* p. 3.

39. Richard J. Barnet and Ronald E. Mueller, *Global Reach: The Power of the Multinational Corporations* (New York: Simon and Schuster, 1974), pp. 92, 396.

40. *Marketing and Promotion of Infant Formula in the Developing Nations, 1978: Hearing before the Subcommittee on Health and Scientific Research of the Committee on Human Resources, United States Senate, Ninety-fifth Congress, Second Session on Examination on the Advertising, Marketing, Promotion, and Use of Infant Formula in Developing Nations, May 23, 1978* (Washington, D.C.: U.S. Government Printing Office, 1978), pp. 126–128.

41. Morton Mintz, "Infant-Formula Maker Battles Boycotters by Painting Them Red," *Washington Post,* January 4, 1981, p. A2.

42. Herman Nickel, "The Corporation Haters," *Fortune,* June 16, 1980, pp. 126–136.

43. Mintz, "Infant Formula-Maker."

44. *Idem.*

45. Ernest W. Lefever, "Politics and Baby Formula in the Third World," *Wall Street Journal,* January 14, 1981, p. 26.

46. Reinhold Niebuhr, *The World Crisis and American Responsibility: Nine Essays,* Ernest W. Lefever, ed. (New York: Association Press, 1958).

47. Ernest W. Lefever, *Ethics and United States Foreign Policy* (New York: Living Age Books, 1957), pp. 115–116.

48. Ernest W. Lefever, "Morality versus Moralism in Foreign Policy," in *Ethics and World Politics: Four Perspectives,* Ernest W. Lefever, ed. (Baltimore: Johns Hopkins University Press, 1972), pp. 1–20.

49. Ernest W. Lefever, "Limits of the Human Rights Standard," in *Morality and Foreign Policy: A Symposium on President Carter's Stance,* Ernest W. Lefever, ed. (Washington, D.C.: Ethics and Public Policy Center of Georgetown University, 1977), pp. 72–74 at p. 72.

50. Ernest W. Lefever, "Intelligence and the American Ethic," in Ernest W. Lefever and Roy Godson, *The CIA and the American Ethic: An Unfinished Debate*

(Washington, D.C.: Ethics and Public Policy Center of Georgetown University, 1979), pp. 1–18 at p. 16.

51. Chetley, *Politics of Baby Food*, p. 95.

52. *Ibidem*, pp. 95–96.

53. *Nomination of Ernest W. Lefever: Hearings before the Committee on Foreign Relations, United States Senate, Ninety-seventh Congress, First Session on Nomination of Ernest W. Lefever, to Be Assistant Secretary of State for Human Rights and Humanitarian Affairs, May 18, 19, June 4 and 5, 1981* (Washington, D.C.: U.S. Government Printing Office, 1981), pp. 499–501.

CHAPTER XI
PRIVATIZING THE COMMON INHERITANCE OF HUMANKIND

1. Robert Gilpin with the assistance of Jean M. Gilpin, *The Political Economy of International Relations* (Princeton, N.J.: Princeton University Press), pp. 31–34.

2. Susan Okie, "Cigarette Ads, Smoking Rise Abroad Linked: Foreign Countries Lowering Barriers to U.S. Tobacco," *Washington Post*, May 5, 1990, p. A3.

3. Congress of the United States Office of Technology Assessment, *New Developments in Biotechnology: Patenting Life—Special Report, OTA-BA-370* (Washington, D.C.: U.S. Government Printing Office, April 1989), p. 53.

4. Calvin M. Kunin, "Resistance to Antimicrobial Drugs—A Worldwide Calamity," *Annals of Internal Medicine* 118, no. 7 (April 1, 1993): 557–561.

5. *Ibidem*, pp. 7, 31.

6. "Application of Ananda M. Chakrabarty, Patent Appeal no. 77–535. United States Court of Customs and Patent Appeals. March 2, 1978," in *Federal Reporter, Second Series, Volume 571 F.2d* (St. Paul, Minn.: West Publishing Co., 1978), pp. 40–47 at pp. 40–43.

7. *Ibidem*, pp. 40–47 at pp. 40–44.

8. "Application of Malcolm E. Bergy, John H. Coats, and Vedpal S. Malik. Application of Ananda M. Chakrabarty. Appeal Nos. 76–712, 77–535. United States Court of Customs and Patent Appeals. March 29, 1979," in *Federal Reporter, Second Series, Volume 596 F.2d* (St. Paul, Minn.: West Publishing Co., 1979), pp. 952–1002 at p. 984.

9. "Diamond, Commissioner of Patents and Trademarks v. Chakrabarty, Certiorari to the United States Court of Customs and Patent Appeals, no. 79–136. Argued March 17, 1980—Decided June 16, 1980," in *United States Reports, Volume 447, Cases Adjudged in the Supreme Court at October Term, 1979* (Washington, D.C.: U.S. Government Printing Office, 1982), pp. 303–322 at p. 303.

10. *Ibidem*, pp. 303–322 at pp. 316–318.

11. *Ibidem*, pp. 303–322 at p. 303.

12. Congress of the United States Office of Technology Assessment, *New Developments in Biotechnology: Patenting Life*, p. 8.

13. M. C., "Animals Can Be Patented," *Science* 236 (April 10, 1987): 144.

14. Leslie Roberts, "Who Owns the Human Genome? Questions are mounting about whether anyone can 'own' the human genome—whether it can be copyrighted or patented—and what effect that might have on a federal collaboration," *Science* 237 (July 24, 1987): 358–361.

15. Marjorie Sun, "Part of AIDS Virus Is Patented," *Science* 239 (February 26, 1988): 970.

16. "Diamond, Commissioner of Patents and Trademarks v. Chakrabarty . . . ," pp. 303–322 at p. 309.

17. Sun, "Part of AIDS Virus Is Patented."

18. Malcolm Gladwell, "Harvard Scientists Win Patent for Genetically Altered Mouse; Award Is First to Be Issued for an Animal," *Washington Post*, April 12, 1988, p. A1.

19. Congress of the United States Office of Technology Assessment, *New Developments in Biotechnology: Patenting Life*, p. 99.

20. Martin Kenney, *Biotechnology: The University-Industrial Complex* (New Haven, Conn.: Yale University Press, 1986), pp. 64–65.

21. Congress of the United States Office of Technology Assessment, *New Developments in Biotechnology: Patenting Life*, p. 121.

22. Mark Crawford, "Patent Claim Buildup Haunts Biotechnology; The backlog of unprocessed patent applications grows by the day as the patent office struggles with limited resources," *Science* 239 (February 12, 1988): 723.

23. S.-C. Lo *et alii*, "A Novel Virus-like Infectious Agent in Patients with AIDS," *American Journal of Tropical Medicine and Hygiene* 40 (1989): 213.

24. William Booth, "AIDS Researchers Upset by Refusal to Share Probes on Mysterious Microbe," *Science* 244 (April 28, 1989): 416.

25. Malcolm W. Browne, "Researchers Enlist Bacteria to Do Battle with Oil Spill; Experiment in Alaska Could Be Landmark in Use of Microbes Against Toxic Chemicals," *New York Times*, May 23, 1989, pp. C1, C13.

26. Edmund L. Andrews, "U.S. Resumes Granting Patents on Genetically Altered Animals; Emphasis Is Now on Medicine Rather Than Food," *New York Times*, February 3, 1993, pp. A1, C5.

27. W. Arthur Lewis, *The Evolution of the International Economic Order* (Princeton, N.J.: Princeton University Press, 1978), pp. 14–46.

28. Gilpin, *Political Economy*, pp. 296–301.

29. Stephen D. Krasner, *Structural Conflict: The Third World Against Global Liberalism* (Berkeley: University of California Press, 1985), pp. 227–250.

30. Robert M. Solow, "Technical Change and the Aggregate Production Function," *Review of Economics and Statistics* 39, no. 3 (August, 1957): 312–320; Eliot Marshall, "Nobel Prize for Theory of Economic Growth; Thirty years ago, Solow proved that technology, not capital, is the key factor in making economies grow—an insight now taken for granted," *Science* 238 (November 6, 1987): 754–755.

31. Marjorie Sun, "UN Biotechnology Center Mired in Politics," *Science* 231 (February 28, 1986): 915.

32. David Swinbanks, "Japanese AIDS Scandal over Trials and Marketing of Coagulants," *Nature* 331, no. 6157 (February 18, 1988): 552.

33. Marlise Simons, "3 Ex-French Officials Convicted in Use of Blood Tainted by H.I.V.," *New York Times*, October 24, 1992, pp. 1, 5.

34. *H.R. 4502 and S. 1966, The Biotechnology Competitiveness Act*, p. 68.

35. *Ibidem*, pp. 57–58.

36. *Ibidem*, pp. 14, 54–55.

37. *Ibidem*, p. 252.

38. *Ibidem*, p. 2.

39. *Ibidem*, p. 16.

40. *Ibidem*, pp. 17, 57, 69, 171.

41. *Ibidem*, p. 15.

42. *Backlog of Patent Applications at the U.S. Patent and Trademark Office and Its Effect on Small High-technology Firms: Hearing before the Subcommittee on Regulation and Business Opportunities of the Committee on Small Business, House of Representatives, One Hundredth Congress, Second Session, Washington, DC, March 29, 1988*, Serial No. 100–59 (Washington, D.C.: U.S. Government Printing Office, 1988).

43. *H.R. 4502 and S. 1966, The Biotechnology Competitiveness Act*, p. 70.

44. *Ibidem*, pp. 39–40, 71, 244.

45. *Ibidem*, p. 217.

46. *Ibidem*, pp. 118–120.

47. *Ibidem*, pp. 109–110.

48. *Ibidem*, p. 225.

49. *Ibidem*, p. 221.

50. *Ibidem*, pp. 198–212.

51. Kenney, *Biotechnology*, pp. 138–141.

52. "Diamond, Commissioner of Patents and Trademarks v. Chakrabarty . . . ," pp. 303–322 at p. 304, asterisked note.

53. Edmund L. Andrews, "Biotechnology Dispute Is Rekindled," *New York Times*, August 31, 1988, pp. D1, D7.

54. *H.R. 4502 and S. 1966, The Biotechnology Competitiveness Act*, p. 176.

55. *Ibidem*, p. 15.

56. *Ibidem*, pp. 177–178.

57. *Ibidem*, pp. 178–179.

58. *Ibidem*, p. 179.

59. David Dickson, "Human Frontiers at the Economic Summit; The Western leaders took note of a Japanese proposal for an international program of biological research, but they stopped short of giving it the go-ahead," *Science* 236 (June 19, 1987): 1518.

60. *H.R. 4502 and S. 1966, The Biotechnology Competitiveness Act*, pp. 179–181.

61. *Ibidem*, pp. 52, 93, 150, 168–169.

62. Leslie Roberts, "NIH Gene Patents, Round Two: NIH has laid claim to 2375 more genes, and critics fear that, [NIH] Director [Bernadine] Healy's protestations aside, NIH is committed to this controversial patenting strategy," *Science* 255, no. 5047 (February 21, 1992): 912–913.

CHAPTER XII
ADVANCING THE PUBLIC HEALTH

1. "Girl's [Jessica McClure's] Rescue Bolsters Texas City's Pride," *New York Times*, October 19, 1987, p. A16; "Tot's [Jessica McClure's] Rescue Wins CNN Highest Ratings Ever," *Atlanta Constitution*, October 22, 1987, p. C10; Viviana A. Zelizer, *Pricing the Priceless Child: The Changing Social Value of Children* (New York: Basic Books, 1985), pp. 138–168.

2. "Unlikely Allies Rush to Free 3 Whales," *New York Times*, October 18, 1988, p. A18.

3. James A. Tobey, *Public Health Law*, 2d ed. (New York: Commonwealth Fund, 1939), pp. 38–56.

4. Thomas Aquinas, *On Princely Government: To the King of Cyprus*, pp. 49–51.

5. Hobbes, *Leviathan*, part 1, chapter 10, pp. 151–152.

6. Sir William Petty, *Political Arithmetick* [1676] in *The Economic Writings of Sir William Petty Together with the Observations upon the Bills of Mortality More Probably by Captain John Graunt*, Charles Henry Hull, ed. (Cambridge, U.K.: Cambridge University Press, 1899), pp. 232–313 at p. 305.

7. *Ibidem*, p. 267.

8. *Ibidem*, p. 259.

9. *Ibidem*, p. 267.

10. Robert D. Willig, "Consumer's Surplus without Apology," *American Economic Review* 66, no. 4 September, 1976; 66(4):589–597.

11. William Stanley Jevons, *The Theory of Political Economy* (1871), 4th ed. (London: Macmillan, 1931), presented in facsimile by Ibis Publishing (Charlottesville, Virginia), p. 14.

12. William Stanley Jevons, "Inaugural Address As President of the Manchester Statistical Society on the Work of the Society in Connection with the Questions of the Day" (November 10, 1869), in *Methods of Social Reform and Other Papers* (London: Macmillan, 1904), pp. 174–186 at pp. 182–184.

13. Universal Declaration of Human Rights, Article 25, Paragraph 1, in *Basic Documents in International Law and World Order*, 2d ed., pp. 298–301 at p. 300.

14. Thomas C. Shevory, "Applying Rawls to Medical Cases: An Investigation into the Usages of Analytical Philosophy," *Journal of Health Politics, Policy and Law* 10, no. 4 (Winter 1986): 749–764.

15. Robert Nozick, *Anarchy, State, and Utopia* (New York: Basic Books, 1974), pp. 232–235.

16. *Ibidem*, p. ix.

17. *Ibidem*, p. 169.

18. H. Tristram Engelhardt, Jr., "Shattuck Lecture—Allocating Scarce Medical Resources and the Availability of Organ Transplantation: Some Moral Presuppositions," *New England Journal of Medicine* 311, no. 1 (July 5, 1984): 66–71.

19. H. Tristram Engelhardt, Jr., *The Foundations of Bioethics* (New York: Oxford University Press, 1986), p. 336.

20. *Ibidem*, pp. 366–367.

21. Robert M. Sade, "Medical Care as a Right: A Refutation," *New England Journal of Medicine* 285, no. 23 (December 2, 1971): 1288–1292.

22. Michael Walzer, *Spheres of Justice: A Defense of Pluralism and Equality* (New York: Basic Books, 1983), pp. 86–91.

23. Larry O. Gostin, Paul D. Cleary, Kenneth H. Mayer, *et alii*, "Screening Immigrants and International Travelers for the Human Immunodeficiency Virus," *New England Journal of Medicine* 322, no. 24 (June 14, 1990): 1743–1746.

24. Kenneth C. Goldberg, Arthur J. Hartz, Steven J. Jacobsen, Henry Krakauer, and Alfred A. Rimm, "Racial and Community Factors Influencing Coronary Artery Bypass Graft Surgery Rates for All 1986 Medicare Patients," *Jour-*

nal of the American Medical Association 267, no. 11 (March 18, 1992): 1473–1477; John Z. Ayanian and Arnold M. Epstein, "Differences in the Use of Procedures between Women and Men Hospitalized for Coronary Heart Disease," *New England Journal of Medicine* 325, no. 4 (July 25, 1991): 221–225; Richard M. Steingart, Milton Packer, Peggy Hamm, *et alii*, "Sex Differences in the Management of Coronary Artery Disease," *New England Journal of Medicine* 325, no. 4 (July 25, 1991): 226–230; Bernadine Healy, "The Yentl Syndrome," *New England Journal of Medicine* 325, no. 4 (July 25, 1991): 274–276.

25. Richard W. Wertz and Dorothy C. Wertz, *Lying-In: A History of Childbirth in America* (New Haven, Conn.: Yale University Press, 1977), pp. 201–233; Kristine Siefert, "An Exemplar of Primary Prevention in Social Work: The Sheppard-Towner Act of 1921," Social Work in Health Care 9, no. 1 (Fall 1983): 87–103; John Duffy, *The Sanitarians: A History of American Public Health* (Urbana: University of Illinois Press, 1990), pp. 239–272; Alice Sardell, "Child Health Policy in the U.S.: The Paradox of Consensus," *Journal of Health Politics, Policy and Law* 15, no. 2 (Summer 1990): 271–304.

26. H. Gilbert Welch and Eric B. Larson, "Dealing with Limited Resources: The Oregon Decision to Curtail Funding for Organ Transplantation," *New England Journal of Medicine* 319, no. 3 (July 21, 1988): 171–173.

27. Glenn Ruffenach, "Health Costs: Transplanting the Cost of Organ Transplants," *Wall Street Journal*, January 2, 1990, p. B1; Edwin Kiester, Jr., "Who Rescued Kay Irwin?" 50 Plus 28, no. 6 (June 1988): 13–14.

28. Ann Japenga, "A Transplant for Coby: Oregon Boy's Death Stirs Debate over State Decision Not to Pay for High-Risk Treatments," *Los Angeles Times*, December 28, 1987, p. V1; Nat Hentoff, "The Rationing of Human Life in Oregon," *Village Voice* 33, no. 21 (May 24, 1988): 44; Ralph M. Crawshaw, Leslie S. Rothenberg, Cory Franklin, and Barney Speight, "Case Studies: That Which Is Wanting . . . ," with commentaries, *Hastings Center Report* 18, no. 6 (December, 1988): 34–37.

29. Marsha F. Goldsmith, "Oregon Pioneers 'More Ethical' Medicaid Coverage with Priority-Setting Project," *Journal of the American Medical Association* 262, no. 2 (July 14, 1989): 176–177; John Kitzhaber, "A Healthier Approach to Health Care," *Issues in Science & Technology* 7, no. 2 (Winter 1990): 59–65.

30. Richard H. Grant, Oregon Health Decisions, written testimony, *Oregon Medicaid Rationing Experiment, Hearing before the Subcommittee on Health and the Environment of the Committee on Energy and Commerce, House of Representatives, One Hundred Second Congress, First Session, September 16, 1991, Serial No. 102–49* (Washington, D.C.: U.S. Government Printing Office, 1991), p. 150.

31. Congress of the United States Office of Technology Assessment, *Evaluation of the Oregon Medicaid Proposal* (Washington, D.C.: U.S. Government Printing Office, 1992), p. 8.

32. Michael J Garland, "Rationing in Public: Oregon's Priority-Setting Methodology," in *Rationing America's Medical Care: The Oregon Plan and Beyond,* Martin A. Strosberg, Joshua M. Weiner, Robert Baker, and I. Alan Fein, eds. (Washington, D.C.: Brookings Institution, 1992), pp. 37–59; David M. Eddy, "What's Going on in Oregon?" *Journal of the American Medical Association* 266, no. 3 (July 17, 1991): 417–420; idem, "Oregon's Methods: Did Cost-Effectiveness Analysis Fail?"

Journal of the American Medical Association 266, no. 15 (October 16, 1991): 2135–2141; idem, "Oregon's Plan: Should It Be Approved?" *Journal of the American Medical Association* 266, no. 17 (November 6, 1991): 2439–2445.

33. Congress of the United States Office of Technology Assessment, *Evaluation of the Oregon Medicaid Proposal*, pp. 15–16.

34. *Oregon Medicaid Rationing Experiment*, pp. 31–161.

35. David C. Hadorn, "Setting Health Care Priorities in Oregon: Cost-effectiveness Meets the Rule of Rescue," *Journal of the American Medical Association* 265, no. 17 (May 1, 1991): 2218–2225; idem, "The Oregon Priority-Setting Exercise: Quality of Life and Public Policy," *Hastings Center Report* 21, no. 3 (May 1991): S11–S16; Jennifer Dixon and H. Gilbert Welch, "Priority Setting: Lessons from Oregon," *Lancet* 337, no. 8746 (April 13, 1991): 891–894; Norman Daniels, "Is the Oregon Rationing Plan Fair?" *Journal of the American Medical Association* 265, no. 17 (May 1, 1991): 2232–2235; William B. Stason, "Oregon's Bold Medicaid Initiative," *Journal of the American Medical Association* 265, no. 17 (May 1, 1991): 2237–2238; Daniel M. Fox and Howard M. Leichter, "Rationing Care in Oregon: The New Accountability," *Health Affairs* 10 (Summer 1991): 7–27; Lawrence D. Brown, "The National Politics of Oregon's Rationing Plan," *Health Affairs* 10 (Summer 1991): 28–51; Joshua M. Wiener, "Oregon's Plan for Health Care Rationing," *Brookings Review* 10, no. 1 (Winter 1992): 26–31.

36. Sara Rosenbaum, "Mothers and Children Last: The Oregon Medicaid Experiment," *American Journal of Law & Medicine* 18, nos. 1–2 (1992): 97–126.

37. Robert Pear, "Plan to Ration Health Care Is Rejected by Government," *New York Times*, August 4, 1992, p. A8.

38. "The Oregon Decision" [editorial], *Washington Post*, August 6, 1992, p. A24; Bob Kuttner, "A Fatally Flawed Plan for Medical Care," *Boston Globe*, August 10, 1992, p. 13.

39. "Washington Prevails" [editorial], *Wall Street Journal*, August 5, 1992, p. A14; "A Bold Medical Plan, Derailed" [editorial], *New York Times*, August 6, 1992, p. A24; Kevin Concannon [Director, Oregon Department of Human Resources], "Dousing a Point of Light: How the Bush Administration Played Politics with the Health of Poor Oregonians," *Washington Post*, August 9, 1992, p. C7; "Give Oregon Health Plan a Chance" [editorial], *New York Times*, August 22, 1992, p. L20; Paul Cotton, "'Basic Benefits' Have Many Variations, Tend to Become Political Issues," *Journal of the American Medical Association* 268, no. 16 (October 28, 1992): 2139–2141.

40. Barbara Roberts [Governor, State of Oregon], "Bush Blows It on Health Care," *New York Times*, August 11, 1992, p. A19.

41. Robert Pear, "White House Expected to Back Oregon's Health-Care Rationing: Plan Covers More People but Fewer Treatments," *New York Times*, March 18, 1993, pp. A1, A10; Daniel M. Fox and Howard M. Leichter, "The Ups and Downs of Oregon's Rationing Plan," *Health Affairs*, Summer 1993, pp. 66–70.

42. Elliott S. Fisher, H. Gilbert Welch, and John E. Wennberg, "Prioritizing Oregon's Hospital Resources: An Example Based on Variations in Discretionary Medical Utilization," *Journal of the American Medical Association*, 267, no. 14 (April 8, 1992): 1925–1931; Carl J. Schramm, "Oregon: A Better Method to Reallocate

Resources?" [editorial], *Journal of the American Medical Association* 267, no. 14 (April 8, 1992): 1967.

43. Michael S. Dukakis, "The States and Health Care Reform" [editorial], *New England Journal of Medicine* 327, no. 15 (October 8, 1992): 1090–1092.

44. *Oregonian* [Portland, Oregon], December 7, 1990; *Oregon Statesman-Journal* [Salem, Oregon], March 17, 1991; *Oregonian* [Portland, Oregon], July 1, 1991; *Oregonian* [Portland, Oregon], July 28, 1991; *Oregon Statesman-Journal* [Salem, Oregon], September 8, 1991; *Register Guard* [Eugene, Oregon], December 3, 1991; *Oregonian* [Portland, Oregon], March 16, 1992.

45. Bernard R. Blishen, *Doctors and Doctrines: The Ideology of Medical Care in Canada* (Toronto: University of Toronto Press, 1969), pp. 116–139.

INDEX